Introduction to
GEOGRAPHICAL HYDROLOGY

SPATIAL ASPECTS OF THE INTERACTIONS BETWEEN
WATER OCCURRENCE AND HUMAN ACTIVITY

EDITED BY

Richard J Chorley

CONTRIBUTORS
R. P. Beckinsale, R. J. Chorley, R. W. Kates, A. V. Kirkby,
R. J. More, M. A. Morgan, R. L. Nace, T. O'Riordan,
J. F. Rooney, W. R. D. Sewell, M. Simons, and C. T. Smith

METHUEN & CO LTD

First published in 1969
First published as a University Paperback in 1971
Reprinted 1974 and 1977

© 1969 Methuen & Co Ltd

Printed in Great Britain by
Richard Clay (The Chaucer Press), Ltd
Bungay, Suffolk

ISBN 0 416 68830 6

Distributed in the USA by
HARPER & ROW PUBLISHERS, INC
BARNES & NOBLE IMPORT DIVISION

Contents

Preface to the Paperback Edition

This paperback originally formed part of a larger, composite volume entitled *Water, Earth, and Man* (Methuen and Co Ltd, London, 1969, 588 pp.), the purpose of which was to provide a synthesis of hydrology, geomorphology, and socio-economic geography. The present book is one of a series of three paperbacks, published simultaneously, which set out these themes separately under the respective titles:

Introduction to Physical Hydrology
Introduction to Fluvial Processes
Introduction to Geographical Hydrology.

The link with the parent volume is maintained by the retention of the Introduction, which gives the rationale for associating the three themes. The aim of this paperback is primarily to make available in a cheap and handy form one of these systematic themes. In doing so, however, it is hoped that the book will provide a constant reminder of the advantages inherent in adopting a unified view of the earth and social sciences, and, in particular, that the study of water in the widest sense presents one of the most logical means of increasing our understanding of the interlocking physical and social environments.

Acknowledgements

The editor and contributors would like to thank the following editors, publishers, and individuals for permission to reproduce figures and tables:

Editors

Canadian Geographer for fig. 7.III(ii).3; *International Science and Technology* for fig. 6.III.2.

Publishers

Heinemann, London, for fig. 4.III(i).1 rom *The Plant in Relation to Water* by R. O. Knight; Macmillan, London, for fig. 12.I.5 from *Techniques for Design of Water Resource Systems* by M. M. Hufschmidt and M. B. Fiering; The Macmillan Co., New York for fig. 4.III(i).5 from *The Nature and Properties of Soils* by H. O. Buckman and N. C. Brady; Princeton University Press for fig. 1.II.1 from *The Quaternary of the United States* by H. E. Wright and D. G. Frey (Eds.); University of Chicago Press for figs. 12.I.13 and 12.I.4 from *Readings in Resource Management and Conservation* by I. Burton and R. W. Kates (Eds.); John Wiley and Sons, Inc., New York, for fig. 4.III(i).3 from *Irrigation Principles and Practices* by O. W. Israelson and V. E. Hansen, and fig. 4.III(i).4 from *Soil and Water Conservation Engineering* by G. O. Schwab *et al.*

Individuals

Chief Engineer, U.S. Bureau of Reclamation for figs. 3.III.2 and 3.III.3; The Controller, Her Majesty's Stationery Office (Crown Copyright Reserved) for fig. 12.I.7; The Director, Irrigation Districts Association of California for fig. 12.I.8; The Director, Road Research Laboratory, Watford for fig. 5.III.5; Ralph Parsons Co., New York, for fig. 11.III.2 of the N.A.W.A.P.A. Scheme; Superintendent of Documents, U.S. Government Printing Office for fig. 7.III(i).1.

Finally, the following thanks are also due:

Members of the City Engineer's Department, Corporation of Bristol for valuable assistance with Chapter 5.III; Mr R. W. Robertson of the University of Victoria for drawing the figures for Chapter 9.III; Mr M. Young, Miss R. King, and

M. J. Ampleford of the Drawing Office, Department of Geography, Cambridge University, for drawing figures for Chapters 3.III, 4.III(i), and 12.I.

The Editor and Publishers would like to thank Mrs D. M. Beckinsale for her painstaking and authoritative preparation of the Index, which has contributed greatly to the value of this volume.

Introduction

R. J. CHORLEY and R. W. KATES

Department of Geography, Cambridge University and Graduate School of Geography, Clark University

Who would not choose to follow the sound of running waters? Its attraction for the normal man is of a natural sympathetic sort. For man is water's child, nine-tenths of our body consists of it, and at a certain stage the foetus possesses gills. For my part I freely admit that the sight of water in whatever form or shape is my most lively and immediate form of natural enjoyment: yes, I would even say that only in contemplation of it do I achieve true self-forgetfulness and feel my own limited individuality merge into the universal.

(Thomas Mann: *Man and his Dog*)

1. 'Physical' and 'human' geography

Perhaps it is of the nature of scholarship that all scholars should think themselves to be living at a time of intellectual revolution. Judged on the basis of the references which they have cited (Stoddart, 1967, pp. 12–13), geographers have long had the impression that they were the immediate heirs of a surge of worthwhile and quotable research. There is good reason to suppose, however, that geography has just passed through a major revolution (Burton, 1963), one of the features of which has been profoundly to affect the traditional relationships between 'physical' and 'human' geography.

Ever since the end of the Second World War drastic changes have been going on in those disciplines which compose physical geography. This has been especially apparent in geomorphology (Chorley, 1965a), where these changes have had the general effect of focusing attention on the relationships between process and form, as distinct from the development of landforms through time. In the early 1950s geomorphologists, especially in Britain, were able to look patronizingly at the social and economic branches of geography and dismiss them as non-scientific, poorly organized, slowly developing, starved of research facilities, dealing with subject matter not amenable to precise statement, and denied the powerful tool of experimentation (Wooldridge and East, 1951, pp. 39–40). It is true that by this time most geographers had long rejected the dictum that physical geography 'controlled' human geography, but most orthodox practitioners at least paid lip service to the idea that there was a physical *basis* to the subject. This view was retained even though traditional geomorphology had little or nothing to contribute to the increasingly urban and industrial preoccupations of human geographers (Chorley, 1965b, p. 35), and its place in the subject as a whole was maintained either as a conditioned reflex or as increasingly embarrassing grafts on to new geographical shoots. American geographers, who had largely abandoned geomorphology to the geologists even before the war, tended to look more to climatology for their physical basis. However, despite the important researches of Thornthwaite and of more recent work exemplified by that of Curry [1952] and Hewes [1965], the proportion of articles relating to weather and climate appearing in major American geographical journals fell more or less steadily from some 37% in 1916 to less than 5% in 1967 (Sewell, Kates, and Phillips, 1968). Even in the middle of the last decade Leighly (1955, p. 317) was drawing attention to the paradox that instructors in physical geography might be required to teach material quite unrelated to their normal objects of research.

The problems of the relationships between physical and human geography facing Leighly were small, however, compared with those which confront us today. Little more than a decade has been sufficient to transform the leading edge of human geography into a 'scientific subject', equipped with all the quantitative and statistical tools the possession of which had previously given some physical geographers such feelings of superiority. Today human geography is not directed towards some unique areally-demarcated assemblage of information which can be viewed either as a mystical *gestalt* expressive of some 'regional personality' or simply as half-digested trivia, depending on one's viewpoint. In contrast, most of the more attractive current work in human geography is aimed at more limited and intellectually viable syntheses of the pattern of human activity over space possessing physical inhomogeneities, leading to the disentangling of universal generalizations from local 'noise' (Haggett, 1965). Today it is human geography which seems to be moving ahead faster, to have the more stimulating intellectual challenges, and to be directing the more imaginative quantitative techniques to their solution.

One immediate result of this revolution has been the demonstration, if this were further needed, that the whole of geomorphology and climatology is not coincident with physical geography, and that the professional aims of the former are quite distinct from those of the latter. This drawing apart of traditional physical and human geography has permitted their needs and distinctions, which had previously been obscure, to emerge more clearly. Perhaps the distinctions may have become too stark, as evidenced by current geographical preoccupations with a rootless regional science and with socio-economic games played out on featureless plains or within the urban sprawl. Perhaps this is what the future holds for geography, but it is clear that without some dialogue between man and the physical environment within a spatial context geography will cease to exist as a discipline.

There is no doubt that the major branches of what was previously called physical geography can exist, and in some cases already are existing, under the umbrella of the earth sciences, quite happily outside geography, and that they are probably the better for it. It is also possible that this will be better for geography in the long run, despite the relevance to it of many of the data and certain of the techniques and philosophical attitudes of the earth sciences. In their place a more meaningful and relevant physical geography may emerge as the product of a new generation of physical geographers who are willing and able to face up to the contemporary needs of the whole subject, and who are prepared to concentrate on the areas of physical reality which are especially relevant to the modern man-oriented geography. It is in the extinction of the traditional division between physical and human geography that new types of collaborative synthesis can arise. Such collaborations will undoubtedly come about in a number of ways, the existence of some of which is already a reality. One way is to take a philosophical attitude implied by an integrated body of techniques or models (commonly spatially oriented) and demonstrate their analogous application to both human and physical phenomena (Woldenberg and Berry, 1967; Haggett and Chorley, In press). Another way is to assume that the stuff of the physical world with which geographers are concerned are its resources – resources in the widest sense; not just coal and iron, but water, ease of movement, and even available space itself. In one sense the present volume represents both these approaches to integration by its concentration on the physical resource of water in all its spatial and temporal inequalities of occurrence, and by its conceptualization of the many systems subsumed under the hydrological cycle (Kates, 1967). In the development of water as a focus of geographical interest the evolution of a human-oriented physical geography and an environmentally sensitive human geography closely related to resource management is well under way.

2. Water as a focus of geographical interest

Water, Earth, and Man, both in organization and content, reflects the foregoing attitudes by illustrating the advantages inherent in adopting a unified view of the earth and social sciences. The theme of this book is that the study of water provides a logical link between an understanding of physical and social environments. Each chapter develops this theme by proceeding from the many aspects of water occurrence to a deeper understanding of natural environments and their fusion with the activities of man in society. In this way water is viewed as a highly variable and mobile resource in the widest sense. Not only is it a commodity which is directly used by man but it is often the mainspring for extensive economic development, commonly an essential element in man's aesthetic experience, and always a major formative factor of the physical and biological environment which provides the stage for his activities. The reader of this volume is thus confronted by one of the great systems of the natural world, the hydrologic cycle, following water through its myriad paths and assessing its impact on earth and man. The hydrologic cycle is a great natural system, but it should become apparent that it is increasingly a technological and social system as well. It has been estimated that 10% of the national wealth of the United States is found in capital structures designed to alter the hydrologic cycle: to collect, divert, and store about a quarter of the available surface water, distribute it where needed, cleanse it, carry it away, and return it to the natural system. The technical structures are omnipresent: dams, reservoirs, aqueducts, canals, tanks, and sewers, and they become increasingly sophisticated in the form of reclamation plants, cooling towers, or nuclear desalinization plants. The social and political system is also pervasive and equally complex, when one reflects on the number of major decision makers involved in the allocation and use of the water resources. White has estimated that for the United States the major decision makers involved in the allocation and use of water include at least 3,700,000 farmers, and the managers of 8,700 irrigation districts, 8,400 drainage districts, 1,600 hydroelectric power plants, 18,100 municipal water-supply systems, 7,700 industrial water-supply systems, 11,400 municipal sewer systems, and 6,600 industrial-waste disposal systems.

This coming together of natural potential and of human need and aspiration provides a unique focus for geographic study. In no other major area of geographic concern has there been such a coalescence of physical and human geography, nor has there developed a dialogue comparable to that which exists between geographers and the many disciplines interested in water. How these events developed is somewhat speculative. First, there is the hydrologic cycle itself, a natural manifestation of great pervasiveness, power, and beauty, that transcends man's territorial and intellectual boundaries. Equally important is that in the human use of water there is clear acknowledgement of man's dependence on environment. This theme, developed by many great teachers and scholars, (e.g. Ackerman, Barrows, Brunhes, Davis, Gilbert, Lewis, Lvovich, Marts, Powell, Thornthwaite, Tricart, and White), is still an important geographic concern, despite the counter trends previously described. Finally, there is no gainsaying the universal appeal of water itself, arising partly from necessity, but also from myth, symbol, and even primitive instinct.

The emergence of water as a field of study has been paralleled in other fields. In the application of this knowledge to water-resource development, a growing consensus emerges as to what constitutes a proper assessment of such development: the estimation of physical potential, the determination of technical and economic feasibility, and the evaluation of social desirability. For each of these there exists a body of standard techniques, new methods of analysis still undergoing development, and a roster of difficult and unsolved problems. Geographers have made varying contributions to these questions, and White reviewed them in 1963. Five years later, what appear to be the major geographic concerns in each area?

Under the heading of resource estimates, White cites two types of estimates of physical potential with particular geographic significance. The first is 'the generalized knowledge of distributions of major resources . . . directly relevant to engineering or social design'. While specific detailed work, he suggests, may be in the province of the pedologist, geologist, or hydrologist, there is urgent need for integrative measures of land and water potential capable of being applied broadly over large areas. The need for such measures has not diminished, but rather would seem enhanced by developments in aerial and satellite reconnaissance that provide new tools of observation, and by the widespread use of computers that provide new capability for data storage and analysis. In the developing world the need is for low-cost appraisal specific to region or project.

A second sort of estimate of potential that calls upon the skills of both the physical and human geographer is to illuminate what White calls 'the problem of the contrast between perception of environment by scientists . . . (and) others who make practical decisions in managing resources of land and water'. These studies of environmental perception have grown rapidly in number, method, and content. They suggest generally that the ways in which water and land resources receive technical appraisal rarely coincide with the appraisals of resource users. This contrast in perception is reflected in turn by the divergence between the planners' or technicians' expectation for development and the actual course of development. There are many concrete examples: the increase in flood damages despite flood-control investment, the almost universal lag in the use of available irrigation water, the widespread rejection of methods of soil conservation and erosion control, and the waves of invasion and retreat into the margins of the arid lands. Thus a geography that seeks to characterize environment as its inhabitants see it provides valued insight for the understanding of resource use.

In 1963 White differentiated between studies of the technology of water management and studies of economic efficiency. Today one can suggest that, increasingly, technical and economic feasibility are seen as related questions. The distinction between these areas, one seen as the province of the engineer and hydrologist, the other as belonging to the economist and economic geographer, is disappearing, encouraged by the impressive results of programmes of collaborative teaching and research between engineering and economics (e.g. at Stanford and Harvard Universities). In this view, the choice of technology and of scale is seen as a problem of cost. The choice of dam site, construction material, and height depends on a comparison of the incremental costs and of the incremental benefits arising from a range of sites, materials, and heights. This decision can be simultaneously related through systems analysis to the potential outputs of the water-resource system.

The methodology for making such determinations has probably outrun our understandings of the actual relationships. The costs and benefits of certain technologies are not always apparent, nor are all the technologies yet known. Geographic research on a broadened range of resource use and specific inquiry into the spatial and ecological linkages (with ensuing costs) of various technologies appears to be required. Indeed, as the new technologies of weather forecasting and modification, desalinization, and cross-basin transport of water and power expand, the need for such study takes on a special urgency.

Finally, there appears to be a growing recognition that much of what may be socially important in assessing the desirability of water-resource development will escape our present techniques of feasibility analysis for much time to come. The need for a wider basis of choice to account for the social desirability of water-resource development persists and deepens as the number of water-related values increase and the means for achieving them multiply. A framework for assessing social desirability still needs devising, but it could be hastened by careful assessment of what actually follows water-resource development. There is much to be learned from the extensive developments planned or already constructed. However, studies such as Wolman's [1967] attempt to measure the impact of dam construction on downstream river morphology or the concerted effort to assess the biological and social changes induced by the man-made lakes in Africa are few and far between. Studies built on the tradition of geographic field research but employing a rigorous research design over an extended period of observation are required. Geographers, freed from the traditional distinction between human and physical geography and with their special sensitivity towards water, earth, and man, have in these both opportunity and challenge.

REFERENCES

ACKERMAN, E. A. [1965], The general relation of technology change to efficiency in water development and water management; In Burton I. and Kates, R., Editors, *Readings in Resource Management and Conservation* (Chicago), pp. 450–67.

BURTON, I. [1963], The quantitative revolution and theoretical geography; *The Canadian Geographer*, 7, 151–62.

BURTON, I. and KATES, R. [1964], The perception of natural hazards in resource management; *Natural Resources Journal*, 3, 412–41.

CHORLEY, R. J. [1965a], The application of quantitative methods to geomorphology; In Chorley, R. J. and Haggett, P., Editors, *Frontiers in Geographical Teaching* (Methuen, London), pp. 147–63.

CHORLEY, R. J. [1965b], A re-evaluation of the geomorphic system of W. M. Davis; In Chorley, R. J. and Haggett, P., Editors, *Frontiers in Geographical Teaching* (Methuen, London), pp. 21–38.

CURRY, L. [1952], Climate and economic life: A new approach with examples from the United States; *Geographical Review*, 42, 367–83.

HAGGETT, P. [1965], *Locational Analysis in Human Geography* (Arnold, London), 339 p.

HAGGETT, P. and CHORLEY, R. J. [1969], *Network Models in Geography* (Arnold, London).

HEWES, L. [1965], Causes of wheat failure in the dry farming region, Central Great Plains, 1939–57; *Economic Geography*, 41, 313–30.

HUFSCHMIDT, M. [1965], The methodology of water-resource system design; In Burton, I. and Kates, R., Editors, *Readings in Resource Management and Conservation* (Chicago), pp. 558–70.

KATES, R. W. [1967], Links between Physical and Human geography; In *Introductory Geography: Viewpoints and Themes* (Washington), pp. 23–31.

LEIGHLY, J. [1955], What has happened to physical geography?; *Annals of the Association of American Geographers*, 45, 309–18.

SEWELL, W. R. D., Editor [1966], Human Dimensions of Weather Modification; *University of Chicago, Department of Geography, Research Paper* 105, 423 p.

SEWELL, W. R. D., KATES, R. W., and PHILLIPS, L. E. [1968], Human response to weather and climate; *Geographical Review*, 58, 262–80.

STODDART, D. R. [1967], Growth and structure of geography; *Transactions of the Institute of British Geographers*, No. 41, 1–19.

WHITE, G. F. [1963], Contribution of geographical analysis to river basin development; *Geographical Journal,* **129,** 412–36.

WHITE, G. F. [1968], *Strategies of American Water Management* (Ann Arbor).

WOLDENBERG, M. J. and BERRY, B. J. L. [1967], Rivers and central places: Analogous systems?; *Journal of Regional Science,* **7** (2), 129–39.

WOLMAN, M. G. [1967], Two problems involving river channel changes and background observations; In Garrison, W. L. and Marble, D. F., Editors, *Quantitative Geography: Part II Physical and Cartographic Topics* (Northwestern University), pp. 67–107.

WOOLDRIDGE, S. W. and EAST, W. G. [1951], *The Spirit and Purpose of Geography* (Hutchinson, London), 176 p.

1.II. World Water Inventory and Control [1]

R. L. NACE
U.S. Geological Survey

The total amount of water in the earth system and its partition and movement among major earth realms have been topics of speculation and investigation during more than a century. Nevertheless, quantitative data are scarce, and the hydrology of the earth as a complete system is still poorly known. Only approximate values can be assigned to most components of the system. Table 1.II.1 is a summary estimate of water in the world exclusive of water of composition and crystallization in rocks and of pore water in sediments beneath the floor of the sea. Most of the water is salty, and much of the fresh water is frozen assets in the cold-storage lockers of Antarctica and Greenland. Most studies of water concern its occurrence and availability in specific areas. However, the global situation has more than intellectual interest, and it is necessary to put local conditions in perspective within that situation.

World globes usually are set up with a prominent land mass to the fore, so the earth looks quite earthy. A more realistic orientation would have Jarvis Island – one of the Line Islands in the Pacific Ocean – in the centre of the field of view. In this perspective the earth is very watery indeed. The preponderance of water area (71% of the earth's surface) and the great extent of the ice-caps are unfortunate in the eyes of people who would prefer more land to accommodate more people (as though the world needed more people!). The effects of reduced ocean area or of melted ice-caps, however, would be far-reaching and generally unfavourable to man.

1. Significance of world inventory

In scientific parlance a system is any region in space together with the things and processes that operate in that region. A closed system is self-contained – nothing enters and nothing leaves. In many ways closed systems are preferable for study to open systems, because once the parameters within the system have been identified their interactions can be studied independently of outside phenomena.

No natural hydrological system is closed. Even the global system is open, because radiant solar energy enters, and reflected and re-radiated energy leaves. However, the solar constant is known more accurately than perhaps any other factor in the hydrological cycle except the physical and chemical properties of water itself. Solar energy is the main driving force of the hydrological cycle.

[1] Publication authorized by the Director, U.S. Geological Survey.

TABLE I.II.I World supply and volume of annually cycled water*

Item	Area (km$^2 \times 10^{-3}$)	Volume (km$^3 \times 10^{-3}$)	% of total water
Atmospheric vapour (water equivalent)	510,000 (at sea-level)	13	0·0001
World ocean	362,033	1,350,400	97·6
Water in land areas:	148,067†	(124,000)‡	—
Rivers (average channel storage)	—	1·7	0·0001
Fresh-water lakes	825	125	0·0094
Saline lakes; inland seas	700	105	0·0076
Soil moisture; vadose water	131,000	150	0·0108
Biological water	131,000	(Negligible)	—
Ground water	131,000	7,000	0·5060
Ice-caps and glaciers	17,000	26,000	1·9250
Total in land areas (rounded)		33,900	2·4590
Total water, all realms (rounded)		1,384,000	100
Cyclic water: Annual evaporation –§			
From world ocean		445	0·0320
From land areas		71	0·0050
Total		516	0·0370
Annual precipitation –			
On world ocean		412	0·0291
On land areas		104	0·0075
Total		516	0·0370
Annual outflow from land to sea – River outflow		29·5	0·0021
Calving, melting, and deflation from ice-caps		2·5	0·0002
Ground-water outflow¶		1·5	0·0001
Total		33·5	0·0024

* Values are approximations, computed on data from many sources which are not mutually consistent. None of the values is precise. Data on evaporation and precipitation modified from L'vovich [1945, p. 54].

† Total land area, including inland waters.

‡ Continental mass above sea-level.

§ Evaporation is a measure of total water participating annually in the water cycle.

¶ Arbitrarily set equal to about 5% of runoff.

Ocean basins, the atmosphere, and the outer crust of the earth form a single gigantic plumbing system, all of whose parts communicate directly or indirectly with all other parts. The occurrence and movement of water in one part of the system are related to its occurrence and movement in all other parts. Furthermore, the whole is more than the sum of its parts, because it includes the interactions of the parts. Therefore the complete system requires study and synoptic observation before men can realistically hope to modify climate or increase water supply predictably and safely. Further, man is the only species which is capable of destroying the habitability of the world. Reckless exploitation of the earth has brought many unwanted side effects that are clearly undesirable and often not understood. It behooves us to understand the entire system in which we live. No other is attainable.

2. Oceanic water

Ocean volume is equivalent to oceanic evaporation during about 3,000 years, which might be taken as the average residence time of a water molecule in the ocean. Some molecules reside only for an instant, however, and water in great ocean deeps may be out of the water cycle during many thousands of years.

Oceanic waters comprise only about 0·023% of the earth's total mass. However, the hydrological cycle is largely an external phenomenon in which oceans have the major role. The system ocean–atmosphere–continents is a great heat engine that drives the water cycle, and oceans are the principal heat reservoir.

3. Water aloft

Evaporation from the huge oceanic reservoir is continual, but an average column of atmosphere contains vapour equivalent only to about 25 mm of liquid water. Estimates of annual precipitation averaged for the whole earth range from 700 to more than 1,000 mm. About 1,000 mm is an acceptable value, equivalent to about 2·7 mm da^{-1} if precipitated uniformly in time and space. The apparent average residence time of a water molecule in the atmosphere is about 10 days, and in any case, water in the atmosphere obviously undergoes rapid flux and reflux.

Strong variability of atmospheric vapour is worth noting. Estimates indicate, for example, 0·6–1·5 mm (water equivalent) over Antarctica (Loewe, 1962, p. 5175) and 50–70 mm in typhoon air masses over Japan (Arakawa, 1959). Using a conservatively rounded value of 50 km hr^{-1} for the typhoon wind speed and 60 mm of water, vapour transport over Japan in a belt 1 km wide in this situation would be equivalent to about 800 m^3 s^{-1}. Considering that major moving air masses are hundreds of kilometres in width, it is apparent that unseen rivers aloft are equivalent to great rivers aground, as they must be to maintain the water cycle.

4. Rivers

The Amazon, mightiest of all rivers, was first seen by Europeans in A.D. 1500. Commanding four caravelles under the Spanish flag, Vicente Yañez Pinzon reached the estuary of the Amazon on 7 February. Four hundred and sixty-three years elapsed before the flow of the river was accurately measured. In 1963–4 a joint Brazilian–United States expedition aboard a Brazilian Navy corvette measured the Amazon's discharge three times, once each at high, low, and intermediate stages.

The average of the Amazon flow at Obidos, 800 km above the mouth (the upstream limit of tidal effects), is 157,000 m³ s⁻¹. The computed average flow at the mouth is 175,000 m³ s⁻¹ (Oltman, 1968). The annual volume of discharge is thus a little more than 5,500 km³, or nearly 20% of the total discharge of all rivers of the world and 4½ times the discharge of the Congo, next largest river.

A. Total discharge

Table 1.11.2 summarizes estimated averaged discharge from all the world's rivers into the sea, amounting to 29,500 km³ yearly. L'vovich [1945, p. 54] estimated a larger total discharge of 36,300 km³, including 1,100 km³ of ice discharge, which is estimated separately herein. Other authors have estimated smaller values, but new data, including that from the International Hydrological Decade, permits improved estimation.

TABLE 1.11.2 Summary of river discharges from land areas to the sea*

Land region	Area† (km² × 10⁻³)	Mean discharge (m³ s⁻¹ × 10⁻³)
Europe, including Iceland	7,960	75·0
Asia and East Indies	31,500	226·0
Africa	18,700	105·4
North America	21,400	151·4
South America	17,000	353·0
Australia, including Tasmania and New Zealand	5,380	13·3
Greenland	2,180	12·4
Totals	104,120	936·5
Equivalent annual volume	29,500 km³	

* Based on data from manuscript (in preparation) by W. H. Durum, R. B. Vice, and S. G. Heidel.
† Includes effective area only; excludes areas that yield no external drainage.

Many hundreds of rivers discharge to the sea, but most of the flow occurs in a few large streams. Sixteen great rivers – those discharging 10,000 m³ s⁻¹ or more – discharge 13,600 km³ annually, or about 45% of the world total. Fifty additional rivers with individual discharges of 500 m³ s⁻¹ or more bring the total to 17,600 km³, or about 60% of all discharge to the sea. Many hundreds of small rivers have not been measured accurately, but individually they contribute little to the total. Aggregate ungauged flow from each continent can be estimated on the basis of climate, topography, vegetation, river characteristics, and other factors. The total discharge estimated in the table is probably within 10% of the true value.

B. Channel storage

Engineers regularly calculate channel storage and storage changes for many river segments as an aid to water management. Although these calculations represent only a small sample of the millions of miles of river channels around the world, they suffice to show that total channel storage, important though it may be locally, is insignificant on the world scale. L'vovich's classical study included an approximate calculation of the isochronal volume of water in river channels. He derived a value of 1,200 km³.

The Amazon channel below Obidos illustrates the relative insignificance of channel storage. Assuming a channel length of 800 km, an average width of 2,500 m, and an average depth of 60 m, channel storage would be about 120 km³, or enough to maintain the flow at the mouth during about eight days.

At median flow the width of the Amazon at Obidos is 2,260 m, and the mean depth is about 46 m (Oltman, 1968). Channel configuration above Obidos is virtually unknown, but it might be assumed that the average cross-section is midway between the dimensions at Obidos and zero at the Peruvian border, 3,000 km distant. If so, upstream storage would be about 160 km³, which brings total storage to 280 km³. This is enough to maintain flow at the mouth during about nineteen days.

Among the hundreds of tributary channels in the vast Amazon river basin, none approach the main stream in capacity. Total storage of 1,000 km³ in the basin might be assumed, equivalent to flow at the mouth during about two months. This is probably excessive.

The Amazon basin covers about 7% of the non-arid, non-glaciated, run-off-producing land area of the world, but other regions do not yield water in the same proportion to area. The Congo basin, equal to 70% of the area of the Amazon, yields only a fourth as much runoff. The Mississippi river basin, half the size of the Amazon basin, yields only a tenth as much runoff. The average daily discharge of all rivers of the world is 80 km³. Inasmuch as 90% of the run-off-producing area is much less copiously supplied with water, and hence with much less channel storage than the Amazon, it may be assumed that total channel storage is 1,700 km³, equivalent to total flow during about three weeks. This value is only a guess, but it illustrates that isochronal channel storage is a vanishingly small percentage of water in the hydrological cycle. At a slow velocity of 1·5 m s⁻¹, in twenty days water could flow from head to mouth of a river 2,600 km in length, so the guess seems reasonable.

The sustained flow of rivers is truly remarkable, considering that precipitation is an unusual event in most areas of the earth. Localization of precipitation in time and space is striking. At Paris, France, a reasonably typical temperate-zone location, total duration of precipitation during a forty-five-year period averaged 577 hours per year (1 year = 8,766 hours), or about 7% of the time (Péguy, 1961, p. 189). Few storms last more than a few hours, so even storm days are mainly rainless. Yet rivers flow throughout the year. The sustaining source of flow is effluent ground water, about which more will be said later.

5. Wide places in rivers

Lakes have been called wide places in rivers. This is true of many small lakes that are impounded by relatively minor and geologically temporary obstructions across river channels. But no single over-simplified metaphor accurately describes all lakes, which vary widely in their physical characteristics and in the geological circumstances under which they occur. The handsome little tarn occupying an ice-scooped basin in the glaciated Alps differs radically from the deep and limpid Crater Lake of Oregon, occupying the collapsed crater of an extinct volcano. The North American Great Lakes occupy huge basins formed in a complex manner by isostatic subsidence of that whole region of the earth's crust, glacial excavation, moraine and outwash deposition, and other factors. These lakes have no resemblance to Lake Tanganyika in the great Rift Valley of Africa, where geological processes created the rift by literally pulling two sections of the earth's crust apart, opening a deep gash, part of which is occupied by the lake. The world contains many other spectacular examples of genetically different lakes.

The earth's land areas are dotted with hundreds of thousands of lakes. Areas like Wisconsin–Minnesota and Finland each contain some tens of thousands. But these small lakes, important though they may be locally, contain only a minor amount of the world supply of fresh surface water, most of which is in a relatively few large lakes on three continents.

The aggregate volume of all fresh-water lakes in the world is about 125,000 km^3, and their surface area is about 825,000 km^2. About 80% of the water is in forty large lakes. For the purpose of this chapter a lake is called large if its contents are 10 km^3 or more. Thus the group excludes water bodies such as the Zürichsee of Switzerland (about 4 km^3). The range of volume among the 'large' lakes is enormous, from the lower limit of about 10 km^3 in Lake Okeechobee (Florida) to an upper limit of nearly 22,000 km^3 in Lake Baikal (central Asia), the world's bulkiest and deepest single body of fresh water. The latter contains nearly as much water as the five Great Lakes of North America. The latter are large in surface area, but their average depth is very much less than that of Baikal.

The Great Lakes and other large lakes in North America contain about 32,000 km^3 of water, which is a fourth of all the liquid fresh surface water in existence. Large lakes of Africa contain 36,000 km^3, or nearly 30% of the earth's total. Asia's Lake Baikal alone contains about 18% of the total.

Lakes on these three continents account for more than 70% of the world's fresh surface water. Large lakes on other continents – Europe, South America, and Australia – contain a comparatively small amount, about 3,000 km^3, or roughly 2% of the total. About a fourth of the total fresh surface-water supply occurs in the hundreds of thousands of rivers and lesser lakes throughout the world.

6. Inland seas

Inland seas and saline lakes of the world are equivalent in volume to fresh-water lakes. Their aggregate area is 700,000 km², and their volume is about 105,000 km³. The distribution, however, is quite different from that of fresh-water lakes. About 80,000 k³ (76% of the total saline volume) is in the Caspian Sea, and most of the remainder is in lesser saline lakes of Asia. North America's shallow Great Salt Lake is relatively insignificant, with less than 30 km³.

7. Life fluid of the vegetable kingdom

Aside from plants that grow directly in water or marshy ground, the great mass of useful vegetation lives on so-called dry land. But the dryness is only relative, and even dust may contain a few per cent of water by weight. Soil holds small amounts of water so tenaciously that plant roots cannot extract it. Desert plants are adapted to low water supply, but most vegetation flourishes only where the soil contains extractable moisture most of the time. A quite ordinary tree may extract and transpire 200 litres of water per day. Frequent replenishment of soil moisture, therefore, is essential for vegetation.

It may be assumed for illustration that the soil zone, except in arid and semi-arid climates, contains on the average about 10% by volume of water within a depth of 2 m beneath the land surface, or about 20 cm of water. In 82×10^6 km² of land area that is not arid to semi-arid and not under permanent ice the estimated total of soil water is about 16,500 km³. Some soil moisture is present in dry areas, and the value actually is larger. From data in an exhaustive study by Van Hylckama [1956], it may be calculated that the average of soil-moisture values for all land areas totals about 25,400 km³. This includes unmelted snow, but snow should be included in the water budget. Therefore Van Hylckama's data are acceptable. The amount of soil water is about fifteen times the amount in channel storage in rivers.

A zone of non-saturation generally occurs between the soil belt and the under-lying water table. By non-saturated flow, water percolates below the root zone and migrates towards the water-table. The amount of migrating water is appreciable even in areas of igneous, metamorphic, and other crystalline rocks. The average thickness of this zone of vadose water is unknown, because average depth to the zone of saturation is indeterminate.

Again, crude estimates indicate the order of magnitude of the water volume. Assuming an average thickness of 15 m for the vadose zone, and assuming that it is perennially wetted at an average field capacity of 20%, the moisture content would be 246,000 km³ in humid to sub-humid areas not covered by ice. This estimate probably is excessive because of the vast areas of crystalline rocks in which the moisture content is much smaller, perhaps on the order of 3–5% as a maximum and much less on the average. It seems reasonable to halve the estimate to obtain an order-of-magnitude value. The assumed rounded value for soil moisture and vadose water together is 150,000 km³.

8. Biological water

Animal bodies are largely water, and plant tissue also contains much water. Plants have an important role in the water cycle through the process of transpiration. In an average temperate area having annual precipitation of 750 mm about 70% may be dissipated by evaporation and transpiration. The two processes cannot be measured separately, and their proportions vary widely in different environments.

Van Hylckama [1956] estimated that water seasonally (ephemerally) stored in vegetation is about $5 \cdot 3 \times 10^{15}$ cm^3 [$= 5 \cdot 3$ km^3]. This does not include water perennially stored in vegetation nor total annual circulation of water through vegetation. Lotka [1956, p. 217] estimated that about 120 mi^3 [500 km^3] of water annually passes through the organic cycle. According to Furon [1967, p. 10], the amount of water needed annually for photosynthesis is about 65×10^{10} tons [$= 650$ km^3]. Both of the latter two values seem unreasonably low.

Precipitation on the world land areas is about 107,000 km^3 annually. If only 25% [26,000 km^3] passes through plants the volume is equal to about 90% of the annual discharge of the world's rivers. Evaporation from land areas is about 71,000 km^3. If only a third of this is vegetative transpiration the annual volume is about 24,000 km^3. This is no indication of the amount of water permanently stored in vegetation, which cannot be large.

9. Unseen reservoirs

The pores in granular geological formations, fissures and joints in hard rocks, and solution cavities and channels in limestone are examples of voids within masses of rock and sediment that can store and transmit water. The storage capacities and transmissivities of rock types range enormously. Storage and transmission are virtually nil in massive dense rock. A clean coarse gravel may have porosity of 25–35%, and rates of flow through the gravel may achieve velocities of 15–20 m da^{-1}. On the other hand, a clay bed may have a porosity of 60%, but will hold water so tenaciously that it moves only in response to osmotic pressure or to compressive stress.

The tremendous volume of ground water in storage is comprehended by relatively few people. Many believe that water occurs underground as lakes and rivers. It is true that limestone caverns contain pools and rivers, but these are exceptional. In general, ground water permeates and fills the pores of underground masses of rock and sediment quite like those we see at the land surface. On the average, ground-water movement is so slow that it would be imperceptible even if the water were visible. The storage capacity of aquifers will be illustrated with a few examples in Chapter 6.III.

Imbeaux [1930, p. 37] cited various estimates of ground-water volume made during the previous hundred years. The smallest of these was 15 km^3 and the largest was 1,175. Even the larger value is absurdly small, because a single major aquifer may contain hundreds of cubic kilometres of water. It is not possible to estimate accurately the total amount of fresh ground water. In many areas deep water is saline, and few regional determinations of the lower limit of fresh water have been made. An order-of-magnitude estimate, however, is permissible and necessary.

Porosity ranges from a small fraction of 1% in massive dense rock to perhaps 35% in highly permeable sediment. With an average effective porosity of only 1%, the upper 1,000 m of the world's land areas exclusive of ice-cap areas (131×10^6 km²) could contain 1.31×10^6 km³ of water. The actual amount is probably at least five times that value, or nearly 7×10^6 km³. Much of this water participates in the hydrological cycle, but an undetermined amount is immobilized in the 9.5×10^6 km² of permafrost area. The residence time of water in aquifers ranges from a few minutes or hours to hundreds of years in most aquifers, but the residence time in some aquifers ranges up to tens of thousands of years.

A great deal of water occurs at depths greater than 1,000 m, but much of it is saline and much is so-called fossil water, not participating in the hydrological cycle. However, it seems safe to round off the estimated volume of recoverable fresh ground water to 7×10^6 km³. Probably an additional equal amount is present but not recoverable for use.

Most circulating ground water enters stream channels within the continents, but some discharges directly into the sea by diffuse percolation and through submarine springs. Submarine springs have been noted in many parts of the world, but few quantitative data on their discharge are available. Owing to low water-table gradients in most coastal areas, the aggregate amount of water thus discharged cannot be a large percentage of cycled water. Crude calculations for the conterminous United States indicate that the amount may be equal to about 5% of streamflow, and this is acceptable as an average value for land areas of the world. The derived total volume is 1,500 km³ per year.

10. Global refrigeration system

The largest share of the earth's fresh water is locked up in the deep-freeze systems of Antarctica and Greenland. Alpine and valley glaciers and small ice-caps are locally important, containing in the aggregate about 80,000 km³ of water equivalent, or about the same amount as in large fresh-water lakes.

The main Greenland ice-cap, nearly 1.73×10^6 km² in area and averaging about 1,500 m in thickness, contains 2.6×10^6 km³ of ice. Outlying ice masses and glaciers covering 60,000 km² increase the total to about 2.65×10^6 km³, equivalent to a water volume of 2.3×10^6 km³. The Antarctic ice-cap, however, is the largest single item in the world's water budget outside the oceans. The area of the main ice sheet in East Antarctica is about 12.1×10^6 km², its average thickness is 2,200 m, and the ice volume is about 26.6×10^6 km³. No good estimate for West Antarctica has been found, but it seems reasonable to round the estimated total to 27×10^6 km³. The equivalent water volume is about 24×10^6 km³. This does not include shelf ice.

The water equivalent of ice-caps and glaciers, in rounded sum, is 26×10^6 km³. These volumes of ice and water are difficult to conceive. The ice-caps and glaciers contain enough water to feed all the rivers of the world at their present discharge rates for nearly 900 years. They could feed the Mississippi River for 46,000 years or the Amazon for somewhat more than a tenth of that time.

A glacier may be regarded as a peculiar kind of river and an ice-cap as a special kind of lake. The great ice-caps and some glaciers discharge water to the sea by melting, by calving, and by wind transport of drifting snow. Only crude data are available, but they seem to indicate that the aggregate water equivalent of discharge by these processes is about 2,500 $km^3 yr^{-1}$.

The average turnover time of water in alpine, piedmont, and valley glaciers is in the range of a few decades to a few centuries. The turnover time for great ice-caps has an extreme range. Today's snowfall at the edge of a cap may return to the sea tomorrow. But snow that forms ice in the heart of the cap may remain in residence during hundreds of thousands of years, and perhaps millions.

In addition to the great ice-caps and lesser glaciers, vast areas in the northern parts of North America, Siberia, and Europe (about 8.5×10^6 km^2) are locked in permafrost – permanently frozen ground. Little of the water in this ground participates in the hydrological cycle. In all, nearly 26×10^6 km^2 of the land area is under ice or otherwise frozen.

11. Possibilities for control

With 97% of the world's water in the sea and 2% in deep freeze, the world evidently is a fine place for whales and penguins, but it has its shortcomings for man. In addition, 17% of the land area is under ice or frozen, and 32% is arid to semi-arid. Small wonder that man, throughout his history, has sought ways to interfere with the water cycle!

Men have managed, mismanaged, and tried with varying degrees of success to control water since the dawn of civilization some 6,000 years ago. Hydraulic engineering has a long and preponderantly successful history. Technology of water control has advanced continually, but the specific means have not changed in hundreds of years, consisting of dams, diversion structures, canals, and the like. The principal change has been in the size and design characteristics of structures and hydraulic systems.

Ever more ambitious plans have now evolved to the stage where armchair planners propose alteration of natural water systems on a continental scale in North and South America and Asia. No doubt engineering skill is equal to the proposed tasks. It is possible also that such projects would be economically feasible from the standpoint of conventional benefit/cost ratios. But engineering and conventional economic feasibility are totally irrelevant in the present state of knowledge about what effects vast projects would have on the ecological systems of the continents.

Human activity has already scarred the face of the earth and upset many ecological systems, but men have not yet learned to predict or forestall the unwanted side effects of landscape alterations. With characteristic heedlessness, however, we continue to plan ever larger alterations, ignoring the fact that construction of monumental projects entails the risk of monumental blunders.

Men have turned their eyes to the skies also and have sought a cloud-wringer to get more water down to earth. Competent analysts say that in at least some situations cloud-seeding has been successful. Completely missing, however, is information whether rain-making at one place diminishes precipitation at another place.

More ambitious is the dream of some men to control world climate. In the present state of knowledge this is beyond possibility. Ability to predict is a good measure of understanding. Weather prediction has not yet advanced from an art to a quantitative science. Until forecasters can make accurate long-range forecasts of natural weather events they can hardly predict the results of efforts at climate modification. It would be foolhardy therefore to try to modify climate, lest the attempt backfire and produce an unwanted change.

Adequate global study of water and weather would be impossible by conventional means, and prohibitively expensive even if possible. Fortunately, new technology offers hope for economical data collection on a global scale and for synoptic communication. So-called remote-sensing techniques are under extensive trial at present. Remote-sensing is any process for determining the nature of an object or phenomenon without direct contact. The human eye is a remote-sensing organ.

Conventional and colour photography with airborne cameras revolutionized geologic and topographic mapping during recent decades. Cameras and other instruments in orbiting or synchronous satellites may, in the future, not only revolutionize mapping but resources surveys in general. Among the more promising techniques are radar and far-infra-red scanning and near-infra-red photography.[1] These have proved useful for mapping topography despite heavy forest cover, for mapping surface drainage, differentiating soils and vegetation, mapping diseased vegetation, detecting geological structures, detecting submarine springs, mapping ocean currents, and a variety of other uses. The whole subject is too complex for treatment in these few paragraphs. The techniques are usable with aeroplanes, but satellites have the advantage that they scan large segments of the earth and its atmosphere during short periods, and they can do this repeatedly.[2]

Acknowledgements. I am grateful to the following colleagues for critical review of the manuscript for this chapter and suggestions for its improvement: P. H. Jones, J. T. Barraclough, J. B. Robertson, and Alfonso Wilson.

[1] See, for example, Colwell, R. N. [1968], Remote sensing of natural resources; *Scientific American*, **21**, No. 1, 54–69.

[2] The United States Government has published an atlas-type volume of remote imagery, consisting of 250 pages in colour. See *Earth Photographs from Gemini III, IV and V* (National Aeronautics and Space Administration SP-129, 1967), $7.00.

REFERENCES

ARAKAWA, H. [1959], Cosmic-ray intensities and liquid-water content in the atmosphere; *Journal of Geophysical Research*, **64**, 625–9.

AUSTRALIAN WATER RESOURCES COUNCIL [1963], *Review of Australia's Water Resources, 1963* (Canberra), 107 p.

CENTRO DE ESTUDIOS HIDROGRÁFICOS [1965], *Anuario de Aforos, 1961–62, v. 9 – Cuencas del Ebro*, Ministerio de Obras Públicas, Madrid [pages not numbered (about 200)].

IMBEAUX, EDOUARD [1930], *Essai d'Hydrologie. Recherche, étude et captage des eaux souterraines* (Paris), 704 p.

LOEWE, F. [1962], On the mass economy of the interior of the Antarctic icecap; *Journal of Geophysical Research*, **67**, 5171–7.

LOTKA, A. J. [1956], *Elements of Mathematical Biology* (New York), 465 p.

L'VOVICH, M. I. [1945], Elementy vodnogo rezhima rek zemnogo shara; *Glavnoe Upravlenie Gidrometeorologischeskoe Sluzhby CCP, Trudy Nauchno – Issledovatel' skikh Uchrezhdenii, Seriya IV, Gidrologiya Sushi*, Bypusk **18**, 109 p.

OLTMAN, R. E. [1968], Reconnaissance investigations of the discharge and water quality of the Amazon; *U.S. Geological Survey Circular 552*.

PÉGUY, C. P. [1961], *Précis de climatologie* (Paris), 347 p.

U.S.S.R. COMMITTEE FOR THE INTERNATIONAL HYDROLOGICAL DECADE [1967], *Water Resources and Water Balance of the Area of the Soviet Union* [in Russian with Russian and English titles] (Hydrometeorological Publishing House, Leningrad), 199 p.

VAN HYLCKAMA, T. E. A. [1956], The water balance of the earth; *Drexel Institute of Technology, Laboratory of Climatology*, Publications in Climatology, **9**, 58–117.

2.III. The Drainage Basin as an Historical Basis for Human Activity

C. T. SMITH

Centre for Latin American Studies, Liverpool University

1. The drainage basin as a unit of historical development

The idea of the drainage basin as a suitable framework for the study and organization of the facts of physical and human geography has a long tradition in the history of the subject. In 1752 Philippe Buache presented a memoir to the French Academy of Sciences in which he outlined the concept of the general topographical unity of the drainage basin. To the cartographers of the late eighteenth and early nineteenth centuries this concept was often caricatured in the mistaken idea that river basins had necessarily to be divided one from another by watersheds so obvious that they could properly be designated by symbols suggesting, in fact, the existence of veritable mountain chains (fig. 2.III.1). Well into the nineteenth century maps sometimes continued to be drawn in this way, even of areas so undistinguished in relief as the plains of European Russia. To some of the academic geographers of the period the identity of the drainage basin, grossly exaggerated as it was by some contemporary cartographers, seemed to offer a concrete and 'natural' unit which could profitably replace political units as the areal context for geographical study (especially in areas of chaotic political fragmentation, such as Germany). It is true that the concept of the watershed as natural boundary and of river systems as unifying networks came into conflict with the idea, also derived to some extent from the study of maps rather than of actual terrain, of the rivers themselves as 'natural' political boundaries, but both concepts disintegrated as knowledge of the earth increased and the work of mapping revealed a much more bewildering complexity (Hartshorne, 1939, p. 45). Ritter, for example, sought refuge in a regional framework which separated upper, middle, and lower portions of stream basins into distinctive types of region (Ritter, 1862), thus reaching a compromise which allowed him to treat of mountain regions in their own right instead of simply as watershed divisions.

Yet in some quarters the idea of the drainage basin as a 'natural' division for regional study survived until quite recently. In his study of the human geography of France, Jean Brunhes based his major divisions of the country on the drainage basins of the Garonne, Loire, Seine, and Rhône–Saône and their major towns. His argument for using this method is based partly on convenience and partly

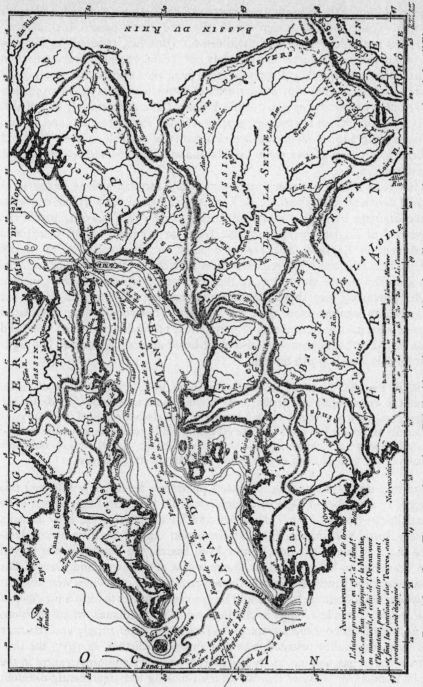

Fig. 2.III.1 Parts of northern France and southern Britain 'dont les eaux s'écoulent directement dans ces mers, depuis les différentes chaînes de montagnes' (From Buache, 1756, Plate XIV).

on a recognition of the importance of water as a link between the earth and man's activities. 'Water is the sovereign wealth of a state and its people. It is nourishment; it is fertilizer; it is power; it is transport' (Brunhes, 1920, p. 93). He goes on to argue that watercourses 'are certainly not the only geographical connection of earth with man, but they do present some principle or possibility of a linkage' (Brunhes, 1920, p. 102). But a more archaic theme is evident in his argument, too, recalling earlier attempts to use the drainage basin as a precisely identifiable areal unit within which to group geographical facts: 'Pays, provinces and regions are not simple natural facts, and are necessarily as variable in their vital and historical expression as they are in their boundaries. Water-courses alone are the precise realities.'

A few years before, in a much more explicit and forward-looking statement of regional divisions of England, C. B. Fawcett also placed considerable emphasis on the drainage basin as a means of territorial division (Fawcett, 1917). His 'provinces' of England were to be the basis for a reorganization of regional government in England, and were arrived at by the application of several criteria. Thus, the size of his regions was to be such that they contained not less than a million people and not more than the total population of the next two smaller provinces; they should contain a regional capital; they should not cut across county loyalties more than was necessary, nor should their boundaries interfere with normal movements of population, particularly between work and home. But their boundaries were also to be based on drainage basins. His argument for this view is worth quoting.

> Since the vital functions of local government include such matters of public health as water-supply and drainage, the making and maintenance of roads and the supply and control of trams, gas and electricity, and since the lines of these are most naturally and easily laid out along the valleys, it will be ordinarily desirable that the boundaries should be drawn near the watersheds, (though) the watershed would only mark out the general trend of a boundary and not govern its details.

In drawing up his map of the proposed provinces, he certainly followed this principle in a number of regions, though not in all. Northern England was to be essentially the basins of the Tyne, Tees, and Eden, bisecting the Lake District and the North York Moors. The Severn region centred on Birmingham consisted more or less of the Severn and Avon basins; the Exe basin was carved from Somerset to be included in the province of Cornwall and Devon. His system was clearly most applicable where watersheds or crest-lines were distinct and where small drainage basins could be grouped together to form a province of suitable size. It was less happy in the Fenland or the Midlands, where divisions had necessarily to be more arbitrary. And although water supply and drainage were frequently valley oriented, one may question whether, in 1917, the provision of trams, gas, and electricity had more than local significance except in the case of London, or whether the cost of road-making was significantly different within drainage basins rather than between them. One is tempted to conclude that Fawcett, like Brunhes to some extent, found the drainage basin and its watershed simply as a convenient means for circumscribing areas which might have been better delineated in terms of urban spheres of influence. Whatever unity had formerly attached to the drainage basin as such in the days of river navigation had largely disappeared in the context of railway-building and the predominance of overland movement. Both Brunhes and Fawcett leave us in doubt as to the precise nature of the link they were seeking to establish, and with the suspicion that such links as there were may have been largely of historical importance by the time they were writing.

Indeed, from the period of primary settlement and colonization to the Industrial Revolution and the revolution in transport, the drainage basin, and more particularly stream networks, had been directly linked with human activity in a variety of ways. Such functional connections could be of two kinds: those which are related to stream networks as water bodies, and those which are related to the existence of systematically arranged patterns of resources (soils, vegetation, local climates, for example), which are themselves organized with respect to the relief, slopes, and stream networks within drainage basins.

The major linkages of the first type would consist of: (a) water supply for domestic purposes and the watering of animals; (b) supplies of fish and game, which are in turn dependent on water supply; (c) a greater ease of accessibility and movement either by navigation or by way of easy overland movement along valley floors; (d) direct water power; (e) water supply for the irrigation of crops. Major linkages of the second type, depending on an association of resources, cannot so easily be generalized, though in areas where the drainage basins are small and relief highly articulated, as in the Alps or in the valleys of peninsular Greece, there may be a complementary pattern of resource which helps to integrate a human society with respect to the drainage basin as a whole. Patterns of seasonal movement between agricultural land in the valleys and highland summer pastures are classic examples of this kind of integration.

Colonization and settlement, particularly before the railway age, have frequently been guided by the lines of accessibility afforded by stream networks, by water supply, or by the concentration of resources afforded by sites in close proximity to the water bodies themselves. In the Anglo-Saxon settlement of England, for example, place-names and archaeological evidence reveal the importance of major river systems as lines of entry and as axes from which subsequent settlement spread towards the watersheds. In the Cambridge region, particularly, early Anglian place-names are found fairly close to the Cam and its tributaries, but the distribution of later clearing and secondary names suggests an expansion of settlement from the valleys towards the watershed regions. The settlement of the East Anglian Stour shows a similar type of sequence, with an added contrast between the nucleated settlement of the valley and the large parishes, dispersed settlement, and late place-names of the watershed. The Welland, the Nene, and the Ouse were similarly lines of entry marked by a scatter of early place-names from which settlement spread towards the watershed zones. In cases such as these easily accessible resources (water supplies, natural meadow, and easily cultivated arable land) may have been more significant than navigability, for the unity of settlement within the drainage basin obviously extends far above navigable limits. And, indeed, primary settlements of the main valleys often continued to retain a leadership in size and wealth long after colonization had been completed (e.g. in the Soar, Nene, Welland, and Cam valleys of the English Midlands).

It would, however, be easy to overemphasize the importance of the drainage basin as such in the primary settlement of north-western Europe, for there were alternative zones of easy movement and attractive resources, such as those offered by the *löss* zone, the relatively open scarp lands of south-eastern England, or the uplands of western Britain.

New links of a different kind were forged in the drainage systems of Western Europe, however, with the emergence of trade and the growth of towns from the tenth and eleventh centuries. In a pre-industrial economy water transport was often more efficient than overland carriage, and for many purposes river navigation was used wherever feasible. Expensive luxuries, such as silks, spices, and dyestuffs, may have been able to stand the cost of transport overland, but bulk trade undoubtedly showed a continued preference for water traffic. The grain trade of the Baltic was channelled along the Oder and the Vistula; timber from central Europe and southern Germany used the waterways; and the expansion and location of viticulture in medieval France and Germany was very strongly controlled by accessibility to navigable water in the Seine basin, the Loire and the Garonne, and in Germany along the Rhine and the Moselle.

Most of the major rivers of western Europe became arteries for the flow of bulk commodities, nourishing a series of urban settlements at their estuaries as seaports, at their lowest feasible bridge-points, at their confluences with major tributaries, at junctions with important overland routes, or at the head of navigation. Indeed, there were very few important towns in western Europe before 1800 which were not located near navigable water. The importance of access to water, even for relatively small towns, may be gauged to some extent from the taxes paid in 1334 by fifty English towns, excluding London. Ten were ports, all on estuaries, and paid an average of £73; twenty-six were on rivers probably used for navigation at this period, and paid an average of £36; ten were not on navigable water, but they paid, on average only £19 in tax; and the remaining four, doubtfully navigable or at best at the head of navigation, paid an average of £14.

It would, indeed, be interesting to test the hypothesis that accessibility by river navigation created a link between the stream networks of a drainage basin and the generation of pre-industrial urban fields, but it would certainly be hazardous to stress too much the dependence of urban settlement and trade on river navigation within the context of individual drainage basins. Long-distance transit trade may have used water transport as much as possible, as for example, between Milan and the Rhine across Italian lakes to the St Gotthard to the Vierwaldstättersee, but the drainage basins they crossed were by no means thereby unified; and man-made obstacles in the form of weirs, taxes, rolls, and brigandage often made overland transport a feasible alternative. And some of the major routes of the Middle Ages, such as the Hellweg across the margins of the Hercynian uplands in Germany, the route from Italy to the Low Countries by way of the Champagne Fairs, or the overland route from Bruges to Cologne, ignored water transport even when it was a viable alternative.

The relative advantage of water routes over land transport has varied greatly, indeed, over time and space. In thickly forested regions difficult of access by any other means water transport has at times provided the only key to exploration, settlement, and exploitation. In the settlement of North America the search for beaver took the fur-traders farther and farther afield to explore the interior of the continent as fur supplies nearer to the Atlantic base were exhausted. For the *coureurs du bois*, living off the country and equipped with the Indian canoe and a few trade goods, the drainage basins were the natural sphere of activity, one linked with another by the shortest possible portages (Brebner, 1933). The short-lived French possession of the Mississippi basin and the foundation of New Orleans were a reflection of this precocious movement. In a similar way, the rapidity of Russian expansion across northern Asia to the Pacific in the seventeenth century also depended on a combination of exploration, the fur trade and the ease of movement along river axes linked by the shortest possible overland portages (Mitchell, 1949). In the Amazon basin the availability of fish and game as well as transport on the major streams supported a more numerous and culturally rich population than the remoter streams and inaccessible watershed regions (Denevan, 1966), and it was precisely these groups who were most strongly affected by the penetration of white populations and cultural change on the opening up of the region by the rubber and cinchona booms of the nineteenth century. In the dense forests of the Amazon basin, the stream network still provides the only integrated transport system, and except in some marginal regions, settlement and commercial activity are almost completely dependent on it.

Changing technologies of transport have, of course, severely eroded such unity as ever attached to the smaller drainage basins of western Europe in terms of navigation and commerce. But in other ways, the coming of steam in the nineteenth century acted as a new stimulus to inland water transport in mid-nineteenth century, for the application of steam power to river navigation from 1807 seemed to emancipate it from the tyranny of sail and variable winds, the tedium of man-power or the restrictions of the towpath. In the Mississippi the great age of the steamboat lasted until the coming of the railway and the re-orientation of the trade of the Mid-West from the south and New Orleans to Chicago and the north-east. In other parts of the world steam navigation seemed to offer the possibility of cheap transport in the opening up of new lands with much smaller capital investment than was needed for railway construction, and ushered in a wave of optimism for the integrated development of major, navigable river basins: the Amazon, the Congo, the Mississippi, the Parana–Paraguay, the Russian rivers, and even the Danube.

Navigation was perhaps the major force creating some sort of unity in the lower parts of drainage basins, but it obviously had no relevance for the upper reaches of shallow waters and high gradients. In a few highly specialized regions of localized industry, these were precisely the zones that achieved some sort of unity through the use of direct water-power for industrial purposes. In the iron industry of the charcoal era, for example, water-power was needed for various purposes, and although the location of iron-making and metal-working industries was usually a product of complex situations and associations of resource, the availability of water-power often controlled the detailed siting of industrial plant, effectively concentrating activity in upland drainage basins. The constraints of water-power were less strong in textile industries before the nineteenth century, though they certainly existed, but waterside locations were needed for wool-washing, fulling, dyeing, and bleaching. Such types of industrial association were to be found in many parts of Western Europe before the Industrial Revolution (Smith, 1967): for example, in the southern tributaries of the Meuse above Liège and Namur (iron and textiles), the Sieg, Lahn, and Wuppertal in Westphalia (iron and textiles), in the minor drainage basins debouching into the St Etienne trough in the Central Massif (metals and textiles), and in various parts of England and Wales (the Yorkshire and Lancashire textile regions, minor valleys above Sheffield, the Stroud valley in Gloucestershire, and in earlier years in the basin of the East Anglian Stour or the Parett basin in Somerset. And where there was navigable water immediately below the zone of exploitable water-power, possibilities existed for the economic integration of a drainage basin on a larger scale, as in the Severn basin or in the Meuse basin below Namur.

2. Irrigation and the unity of the drainage basin

The closest and probably the most widespread association of past human activity with the hydrological balance, relief, slopes, and stream networks of the drainage basin has been achieved through the operation of irrigation systems. Primitive irrigation systems are discussed elsewhere (Chapter 4.III(ii)), but in the context of the drainage basin, the most important types of irrigation are those which involve some degree of communal action for their construction, operation, and maintenance. Of these the chief (in terms of historical importance and geographical spread) are those which use water from a stream network and have more or less elaborate systems of canals for the distribution of water to individual settlements. Irrigation of this type has been characteristic of the great hydraulic societies of India and China or of the New World in pre-Columbian Peru. Limits are set to the extension of irrigation systems of this type by the availability of suitable terrain for cultivation or for the distribution of water, usually within the context of a single drainage basin; by the volume and seasonal regime of water supply; by the technology available; and not least by the scale and nature of social and political organization within which the construction, maintenance, and administration of water-control must be carried out (Forbes, 1965, pp. 4–5). Small-scale and piecemeal irrigation is possible within parts of a drainage basin, and early operations in Western Europe and Peru were often of this kind. But competition for water supply and the pressure of population leading to demands for the extension of irrigation to new land frequently seem to have led at an early stage to the integration of piecemeal systems within the drainage basin as a whole. In coastal Peru, for example, early piecemeal irrigation was soon replaced by canal systems incorporating the whole of the irrigable area of a drainage basin below the Andean foothills. And parallel developments, involving the drainage basin (or the lower part of it) as an integrated unit for defence, religious and administrative control, and the supply of urban centres, point also to the extension of a single political authority over the whole of a valley region (Willey, 1953).

Wittfogel has seen the social and political implications of pre-industrial irrigation systems of water control in terms of coercion and the formation of 'oriental despotism' (Wittfogel, 1957). The relevant part of his argument may be briefly summarized, that the construction and maintenance of large-scale irrigation systems require the assembly of a considerable labour force which may be most efficiently created either by the institution of forced labour or the levy of tribute and taxation or both. A centralized administration is also needed for the maintenance of canals and to control water distribution. The administration in control of the distribution of water is, in effect, in complete control of agricultural activity, and is thus in a position to demand complete authority and complete submissiveness, subject only to mass revolt and rebellion in the face of desperate conditions. Society becomes polarized, in fact, into an illiterate, dependent peasantry and an *élite*, as in the traditional bureaucratic governments of China.

The role of the drainage basin as the fundamental territorial unit on which this type of administrative superstructure may be built and with respect to which the irrigation systems are constructed has not been systematically examined, though it is clearly evident in coastal Peru, where the integration of individual basins was followed by the creation of coastal empires built up from adjacent valley systems, unified by dominant urban centres such as that at Chan-Chan (Bushnell, 1956, p. 114). In China an attempt has been made to establish a relationship between the establishment of new irrigation works and the geographical basis of political power in some of the early Chinese dynasties, in which the unit of the drainage basin has considerable importance (Chao-Ting Chi, 1936). He identified the source of early dynastic power (of the Ch'in and Han dynasties) in the third century B.C. in the zone of the Loess Highlands, where the tributary valleys of the Wei, Fen, Lo, and Chin converge at the great bend of the Hwang Ho. These were small valleys, easily cleared of their woodland and with *löss* soils fertile under irrigation. Streams were relatively short and much more easily controlled than the Hwang Ho itself. The scale of the environment was sufficiently small for the control of these minor drainage basins to be achieved by available technology and administration, but the irrigation systems of these valleys, and particularly the Wei Ho, supported a powerful bureaucratic state with its capital in the Wei valley itself at Chang'an (near Sian). New and larger irrigation works later widened the basis of tribute-collection and thus of political power. They were made possible by new technological advance and by the increase of population, which was itself a measure of the success of earlier irrigation projects. And these new works were now located in the middle course of the Hwang Ho in the North China Plain. Significantly, the capital was shifted to Loyang at the margin of the North China Plain itself, which now replaced the Loess Highlands as the 'key economic area' of Chinese government. Subsequently, however, the shift of irrigation projects farther south into the Hwai River basin, the Red basin of Szechuan, and then the Lower Yangtze created a new basis of economic and political power, and by the time of the T'ang dynasty between the seventh and the tenth centuries A.D. the 'key economic area' had shifted towards central China and the Yangtze basin.

In a variety of ways the drainage basin has formed a framework for human activity: in guiding the direction of primary settlement, in river navigation and the growth of trade and towns, in the provision of water-power for industrial concentrations, and in providing a logical context for irrigation works. But in many cases these functions have created a unity, not in the drainage basin as a whole but rather in those parts of a drainage basin which have relevance for a particular activity. Few recognized, even dimly, the interrelation of the whole drainage basin in any conscious way. The Chinese never appear to have realized, for example, the relationship between deforestation in the Loess Highlands and the floods, silting, and droughts of the North China Plain.

Indeed, the idea of the drainage basin as an appropriate areal unit for the organization of human activity and for regional planning has only recently been revived with the recognition of the basin as an interrelated system in which soil and vegetation cover as well as hydrological balance are involved, and with the recognition of the need for integrated plans and policies to deal with problems posed by flood-control, sedimentation, soil erosion, hydroelectric power production, navigation, and even nature conservation and stream pollution. The Tennessee Valley Authority fulfilled such a need for a regional authority transcending the boundaries of traditional units of government, and it was the prototype from which stemmed other proposals and organizations: a Missouri valley authority; an abortive proposal for a Danube Valley authority, mooted just after the Second World War, when the future of central Europe was still in the melting pot (Kish, 1947); and the regional corporations of the Cauca valley (1954) and the Magdalena valley (1960) in Colombia (Banco de la Republica, 1962). Gilbert White has written of the Mekong valley project as a programme transcending political boundaries, and aiming at an integrated approach to the problem of water control and environmental planning in the greater part of a major drainage basin (White, 1963). And in recent months a programme has been announced for a preliminary study of the feasibility of an integrated development scheme for the basins of the Paraná, Paraguay, Uruguay, and River Plate which will involve co-operation by Argentina, Bolivia, Brazil, Paraguay, and Uruguay (*Peruvian Times*, 1967).

REFERENCES

BANCO DE LA REPUBLICA (Colombia) [1962], *Atlas de Economía Colombiana*; Volume 4, Aspectos agropecuarios y su fundamento ecologico.

BREBNER, J. B. [1933], *Explorers of North America 1492–1806;* (London), 501 p.

BRUNHES, J. [1920], *Géographie humaine de la France*; Hanotaux, 6; Editor, Histoire de la France, Vol. 1 (Paris), 493 p.

BUACHE, M. [1756], Essai de géographie physique; *Mémoires de Mathematique et de Physique, Académie Royale des Sciences*, 1752, pp. 399–416.

BUSHNELL, G. H. S. [1956], *Peru* (London), 216 p.

CHAO-TING CHI, T. [1936], *Key Economic Areas in Chinese History* (London), 168 p.

DENEVAN, W. M. [1966], A cultural-ecological view of former aboriginal settlement in the Amazon basin; *The Professional Geographer*, **18**, 346–51.

FAWCETT, C. B. [1917], The natural divisions of England; *Geographical Journal*, **49**, 124–41.

FORBES, R. J. [1965], *Irrigation and Power*; Studies in Ancient Technology; Vol. 2 (Leiden), 220 p.

HARTSHORNE, R. [1939], *The Nature of Geography* (Pennsylvania), 482 p.

KISH, G. [1947], TVA on the Danube?; *Geographical Review*, **37**, 274–302.

MITCHELL, M. [1949], *The Maritime History of Russia, 848–1948* (London), 530 p.

Peruvian Times [1967], news article, 14 July.

RITTER, K. [1862], *Comparative Geography*; translated W. L. Gage (Edinburgh), 254 p.

SMITH, C. T. [1967], *An Historical Geography of Western Europe before 1800* (London), 582 p.

TECLAFF, L. A. [1967], *The River Basin in History and Law* (Martinus Nidhoff, the Hague), 228 p.

WHITE, G. F. [1963], Contributions of geographical analysis to river basin development; *Geographical Journal*, **129**, 412–36.

WILLEY, G. R. [1953], Prehistoric settlement patterns in the Virú valley; *Bureau of American Ethnology*, Smithsonian Institute, Bulletin 155, 453 p.

WITTFOGEL, K. A. [1957], *Oriental Despotism* (New Haven and London), 556 p.

3.III. The Interaction of Precipitation and Man

R. J. CHORLEY and ROSEMARY J. MORE

Department of Geography, Cambridge University and formerly of Department of Civil Engineering, Imperial College, London University

1. Man's intervention in the hydrological cycle

Every part of the hydrological cycle has been tampered with; runoff is stored behind dams, evaporation is reduced by coating water surfaces with suitable monolayers, transpiration losses are reduced by removing phreatophytes, and ground water is recharged by water spreading and pumping (fig. 3.III.1). It is now clear that, under suitable conditions, natural precipitation can be artificially modified. Even where the actual amount of water cannot be tampered with, man's response to its occurrence is far from passive, in that he changes its circulation by the use of irrigation and, as a last resort, gambles on its occurrence by crop insurance.

Since precipitation is the principal input into the hydrological cycle, man's attempts to modify it will have consequences throughout the working of the cycle. For example, precipitation modification would influence magnitude and time of runoff, soil moisture reserves, and ground-water storage. It is because of these all-pervading climatic, hydrologic, and social repercussions and possible side-effects throughout the hydrological system (some known and many unknown) that man has hesitated to embark upon unbridled precipitation changes, although technologically he is increasingly in a position to be able to make them. However, man is only at the beginning of weather modification, being able to achieve only local effects of small magnitude, so that the wider-scale consequences which surround his endeavours have not yet been encountered.

There are two main approaches to conscious precipitation modification by man, quite apart from the locally important inadvertent effects associated with the construction of large urban and industrial complexes, the creation of large lakes, and the modification of the surface vegetational cover. The first is to attempt to achieve increases of precipitation of a known magnitude (5 or 10%) by available techniques, generally with a view to increasing soil moisture for plants or increased water yield from catchments for public water supply and other uses. An alternative, and broader, viewpoint is to try to assess the ideal water requirement for any location. Although this is difficult to evaluate, involving as it does the determination of the changing uses of the water and reconciliation of competing interests, it is probably a sounder aim than the more limited one of simply increasing precipitation. Thus attention should ideally be

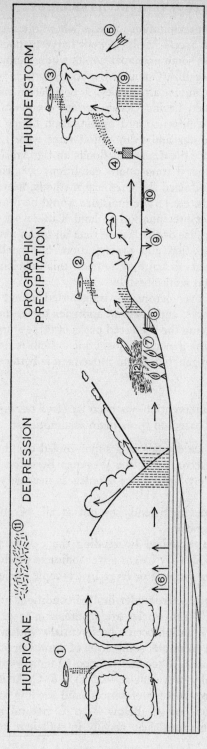

Fig. 3.III.1   Some points of human intervention in the world hydrological cycle.

1. Seeding of hurricane eye-wall.
2. Seeding of orographic cloud.
3. Seeding of thunderstorm.
4. Ground-based silver iodide seeding.
5. Dispelling hail by rocket.
6. Interference with sea-surface evaporation.
7. Irrigation below dam.
8. Artificial reservoir.
9. Water spreading and ground-water recharge.
10. A 'thermal mountain'.
11. Needles in orbit.
12. Local fog dispersion.

given not only to increasing precipitation in areas where there is too little but also to decreasing it where an excess results in poor harvests, flooding and other disasters. Figure 3.III.1 shows some principal points where human intervention in the hydrological cycle is possible. One is at the ocean surface, where moisture rises by evaporation to form moist air masses. A correlation has been shown to exist, for example, between Pacific Ocean surface temperatures 800 miles west of the Sierra Nevada, California, and precipitation falling on this range. Similarly, high sea temperatures and evaporation rates were recorded in the Western Mediterranean before the disastrous floods on the Arno at Florence in 1966. It would seem logical that if atmospheric conditions favouring evaporation from the oceans could be optimized (by chemical methods, heating the surface water, increasing wind speeds, etc.) more moisture would be drawn up into the atmosphere to be available for precipitation on land. Little work has been done on this approach, largely because of the lack of data on ocean temperatures and their correlation with evaporation and precipitation, but satellite sensing of meteorological conditions over oceans may remedy this deficiency in the near future and make experimental work possible.

The saturated air masses over oceans are transported inland largely by depressions following storm tracks, and a second approach to weather modification might be possible by influencing the preferred paths of these storm tracks. Such a large-scale intrusion into the general atmospheric system is not yet possible, but may be more so as storm generation and movement is better understood.

2. Rain-making

The most fruitful point of intervention has been by *cloud seeding*. Rain-making experiments of this type are based on three main assumptions:

1. Either the presence of ice crystals in a super-cooled cloud is necessary to release snow and rain (according to the Wegener–Bergeron theory); or the presence of comparatively large water droplets is necessary to initiate the coalescence process.
2. Some clouds precipitate inefficiently or not at all, because these components are naturally deficient.
3. The deficiency can be remedied by seeding the clouds artificially with either solid carbon dioxide (dry ice) or silver iodide to produce ice crystals, or by introducing water droplets or large hygroscopic nuclei.

Such seeding is thus only productive under limited conditions of orographic lift and in thunderstorm cells, when nuclei are insufficient to generate rain by natural means. Natural precipitation occurs preferentially within certain upper-air temperature ranges – for example, some 80% of winter precipitation in the state of Washington falls when the 700-mb (10,000-ft) temperature is not lower than $-10°$ C and is especially prevalent at $-4°$ to $-8°$ C. Artificial precipitation stimulation must exploit these preferences, and seeding is thus effective within a limited temperature range. Below $-20°$ C natural nuclei, such as dust, become active to form snowflakes, usually in sufficient numbers so that

A. Recognition: radar, balloon, and observing aircraft.

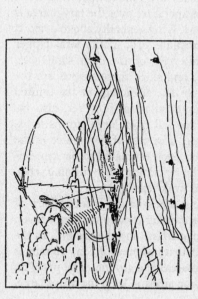

C. Treatment: cloud developing intensively and pre-cipitation falling.

B. Treatment: ground-based silver iodide generator and aircraft dispensing silver iodide or dry ice.

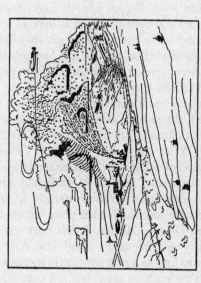

D. Evaluation: precipitation, depth, and cloud informa-tion transmitted to headquarters for evaluation.

Fig. 3.III.2 The operation of cloud seeding (Atmospheric Water Resources Program, 1968).

additional silver iodide particles are not needed, and under some conditions are actually detrimental. Cloud seeding may be effected by burning silver-iodide-impregnated fuels or solutions at ground level to produce a smoke which is carried upwards by wind into the effective zone, by firing rockets containing nuclei into the effective zone, or, more usually, by dropping the nuclei from aircraft (fig. 3.III.2).

Cloud seeding by these means has been attempted in many parts of the world, notably in Australia and the western United States. The need for fresh sources of water now and in the future is so acute in the United States that the Bureau of Reclamation has initiated the nation-wide Project Skywater to investigate the possibilities of water management through artificial rain-making. The purpose of one such scheme is to increase winter precipitation over the mountains of the Upper Colorado River, thus augmenting spring runoff, which would be stored and regulated to meet demand by the existing reservoirs. An increase of the November–April precipitation by 15% over 14,200 square miles of target areas is expected to yield an average additional runoff of 1,870,000 acre-feet annually. An important advantage of water provision by cloud seeding methods is that a 10% increase in rainfall can result in a 17–20% increase in runoff, because evaporation does not increase in proportion to the greater precipitation, so that there is more water available for runoff. Preliminary investigations have shown that the optimum conditions for seeding are when there is a thin (less than 5,000 ft thick) saturated air-mass layer, the temperature at the top of which is not less than −20° C, and the temperature over the target area is less than −10° C (fig. 3.III.3). Eight major areas, lying generally above 9,500 ft where annual natural runoff is over 10 in., contribute 75% of the total Upper Colorado basin runoff, although they form only 13% of the basin land area. Most precipitation comes in a few big storms, and since these storms are the main precipitation generators, it is important to take advantage of the limited opportunities they offer. Increasing the total precipitation, however, also increases the variability of its occurrence, since the fall from large storms is increased, the smaller rainfalls remaining the same. The cost of new water in the Upper Colorado, provided by cloud seeding, has been estimated to be approximately $1.00–$1.50 per acre-foot. (This is the operating and running cost, exclusive of the research necessary to make the work feasible.)

An associated method of artificially increasing precipitation is a corollary of cloud seeding and involves attempts to reduce the water loss by the evaporation of precipitation which takes place between the cloud base and the ground. In the Sonora Desert, Mexico, it has been estimated that, whereas 40 in. of rain is annually available at the cloud base, only 9 in. reaches the ground. The problem here is to find an agent which will increase the drop size and keep the drops large, and so far such an agent has not been found.

In summary, it is clear that there is a limited range of natural conditions in which significant artificial interventions can be made to produce, increase, or conserve precipitation. The seeding of some cumulus clouds at temperatures of about −10° to −15° C probably produces a mean increase of precipitation of

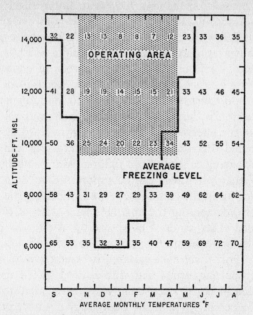

Fig. 3.III.3 Average temperature–altitude chart for Grand Junction, Colorado, showing the average freezing level and the optimum operating area for seeding (i.e. between November and April above 9,500 ft) (Hurley, P. A., 1967, *Augmenting Upper Colorado river basin water supply by weather modification;* Annual Meeting American Society of Civil Engineers, New York, October 18, 1967).

some 10–20% from clouds which are already precipitating or 'are about to precipitate', with comparable increases up to 250 km downwind, and increases of up to 10–15% have resulted from the seeding of winter orographic storms. On the other hand, the seeding of depressions has produced no apparent increases, and it appears that clouds with an abundance of natural nuclei, or with above-freezing temperatures throughout, are not susceptible to rain-making. At present it is often a difficult statistical matter to determine whether many of man's attempts have produced significant increases in precipitation; for example, six experiments in Washington and Oregon produced the following probabilities that rainfall had been increased: 95, 67, 50, 50, 50, and 41%! Another instance serves also to highlight the possible legal problems which attempts at rain-making will provoke. In Quebec a recent 25% increase in rainfall coincided with rain-making attempts, causing extensive floods, crop damage, and disruption of the tourist industry. Following a large public outcry, the Federal Government announced that the effect of the seeding had been to *decrease* the possible rainfall receipt by 5%!

Besides rain-making, other human interventions in precipitation involve the successful local dissipation of freezing fogs over airports by spraying with propane gas, brine, or dry ice, causing snow to fall and clear the air. The Russians have also claimed success in dissipating damaging hailstorms by the use of radar-directed artillery shells and rockets to inject silver iodide into high-liquid-content portions of clouds, which freezes the available super-cooled water and prevents it from accreting as shells on growing ice crystals.

It is probable, however, that man is on the brink of much larger interventions into the hydrological cycle on a scale of hundreds of square miles or more. Hurricanes cause, on average, some $300 million worth of damage annually in the United States and Canada. At present the $9 million spent on forecasting, warning and protection is estimated to save some $25 million of property, with only about 20% of the affected population being involved in protective action. It has been estimated that improvements in forecasting and warning systems might increase the saving to some $100 million. More ambitious projects make it likely, however, that hurricanes can be suppressed or 'damped down' by the seeding of the rising air in the cumulus eye-wall, widening the ring of condensation and updraught, decreasing the angular momentum of the storm and thus the maximum speed of its winds. The spreading of the sea ahead of the storm with oily materials might be used to cut off surface evaporation and thus the hurricane energy supply. Even such apparently beneficial attempts may represent potentially dangerous tampering with the natural global moisture economy, especially so in this instance when it is remembered that 30% of the August rainfall of the Texas coast, 30% of the September rainfall of the Louisiana and Connecticut coasts, and fully 40% of the September rainfall at Atlantic City, New Jersey, are derived from hurricane circulations. Even more speculative schemes involve putting huge quantities of dust or metallic needles into stationary orbit to locally reduce sea temperatures and decrease evaporation; as well as creating 'thermal mountains' by painting desert surfaces black to increase their conservation of solar heat, stimulate convection, and thereby increase cloudiness and precipitation downwind. The unknown dangers attendant upon such large-scale tampering with the delicately balanced world hydrological cycle must postpone such schemes until theoretical mathematical models simulating the behaviour of the earth–atmosphere system have been developed so that all the possible effects can be predicted in advance.

3. Spray irrigation

An alternative method of precipitation modification is to supply the rain artificially at ground level by means of sprinkler equipment. This method is used most commonly for crop irrigation in areas of supplemental irrigation (e.g. North-West Europe, Eastern United States) or in arid and semi-arid areas with a highly capitalized agriculture (e.g. California and Israel). The amount of irrigation water needed is calculated by estimating the potential evapotranspiration of the cropped area by the most appropriate of the available formulae (e.g. by Penman, Lowry and Johnson, Blaney-Criddle; see Chapter 4.1). In many areas a supply of soil moisture is built up in the wet season, and this reserve can be deducted from the calculated potential evapotranspiration, as it is stored water available for the plant's use in the dry season. Calculations of additional water needs for cropping are easier in arid countries than in areas of supplemental irrigation (i.e. in arid areas no rain may be expected), whereas in semi-humid areas calculations have to have a probability factor depending on the assessment of likelihood of growing-season rainfall.

The rain produced by sprinkler irrigation may be called *artificial rain* (in contrast to the natural rain from atmospheric intervention methods), in that it can be controlled, in its time of occurrence, duration, intensity, uniformity, and drop size, by the type of sprinkler equipment chosen and the spacing of the equipment. Precise specifications for artificial rainfall are most closely achieved by sprinklers used in laboratory catchments (see Chapter 2.1). In addition to fixed sprayers and spraylines, more sophisticated methods have been used in the Western United States and in Hawaii, where sprays are mounted on wheels and can be moved by electronic means across the crop, thus simulating the passage of a shower of rain. With this type of equipment evaporative losses can be kept to a minimum, especially if the sprinkling is done at night, at dawn or in the evening (when the sunlight is least), or when the wind speed is low. Winds can seriously interfere with the efficiency of the distribution, particularly from rainguns, although special sprinklers can be used if the prevailing winds are high. Sprinkler irrigation is the most mechanized and probably the most efficient method of applying water to a great range of crops, and the costs are correspondingly higher than those for other methods of irrigation. An interesting elaboration is that similar equipment, with a freezer added, has been used for snow generation in the ski-ing resorts of the Swiss, Austrian, and Italian Alps and in the Cairngorms (Scotland).

4. Gambling with water

While man has made a certain amount of progress in increasing, conserving, and redistributing rainfall, he has done very little to protect himself from the effects of too much. Future architectural advances in wide-arc roofing, already fore-shadowed by the Houston *Astrodome*, may make it possible for urban precincts, and even whole cities, to have controlled climates. Most rural activities, with the limited exception of glasshouse protection for certain high-value crops, must proceed under the threat of the occurrence of droughts, floods, hailstorms, blizzards, and the like, although their probabilities of occurrence vary very much from place to place. Although certain industries, notably building con-struction, transport, and those geared to a minimum level of streamflow, are victims of these meteorological events, it is the farmer who must gamble most heavily on hydrological probabilities. One of the most obvious ways of doing this is to diversify cropping so that, whatever the weather, some minimum level of production is reasonably assured. Such diversification is particularly apparent in regions of subsistence agriculture, where rainfall is highly variable. In East Africa, for example, more than one crop failure in ten can be disastrous where native farmers have no capital reserves, in contrast to which local commercial agriculture finds one rainfall failure (i.e. less than 30 in. of rain per annum) in three 'acceptable'. Game theory has been used to show how an unconscious perception of climatic probabilities has been employed by primitive farmers to maximize their possibility of continual survival against nature by always planting a balanced range of crops with differing responses to deficiency or abundance of rainfall.

Although diversification is also an essential feature of much advanced farming practice, the changing commercial economic base means that in many regions the narrowing profit margins require ever-increasing production per acre, which, in turn, forces increasing specialization and mechanization, both of which imply economic pressures towards monoculture. At the same time the attendant rising capital costs and narrowing profit margins mean that a single year's crop may represent an investment of four or more years' profits. The narrowing of the economic base of regional agriculture makes climatically-controlled variations in agricultural production of great significance, because all over the world agriculture is still largely at the mercy of the weather, despite advances in weather-resistant crops and the use of weather forecasts (Sewell, Kates and Phillips, 1968, p. 267). Many studies have been made regarding the economic effects of weather variations on single regional crops, for example the effect of drought on wheat failure in the central Great Plains (Hewes, 1965), of July rainfall and mean temperatures on corn production in Indiana (Visher, 1940), of rainfall on rice yields in West Bengal (Hore, 1964), and of January precipitation on milk butterfat contents and South Auckland, New Zealand (Maunder, 1968). The last-named study, for example, showing that the occurrence of a wet January (on average 1 in 6) is 'worth' about $2 million (New Zealand) to the dairy farmers of South Auckland when all the benefits have been evaluated. Even under the reasonably-predictable British climatic conditions economically-significant moisture variations are of great importance. For example, the generally wet and cloudy summers of the 1950's retarded plant growth, encouraged diseases and made cereal harvesting difficult or impossible; and the severe snowfalls of 1947 and 1963 caused severe stock losses. Even grass, as in New Zealand, is very sensitive to moisture variations and, under high stocking conditions in south-east England without irrigation, the milk output may fall from 600 gallons per acre in a wet year to 200 gallons in a dry one, turning a profit of £24 per acre into a loss of £16. More sophisticated studies have been concerned with the relationships between climatic studies and farm organization and management programmes (Curry, 1952 and 1962) and with the multivariate effect of many climatic parameters not only on agricultural production but on the whole economy of a region (Maunder, 1966).

One partial answer to such climatically-controlled variations in agricultural returns is crop insurance, and in the United States more than $2·7 billion worth is held by farmers. This covers twenty-three crops, the most widespread being for wheat, maize, soyabeans, cotton, tobacco, oats, and barley (in that order). For some crops (e.g. tobacco) insurance commonly guarantees a stated income per acre, whereas for others (e.g. maize) it guarantees a given production per acre, at a stated quality. Premiums and conditions naturally vary greatly, depending on fertility, as well as climatic and other risks, and in a single United States' county the guaranteed levels of maize production may range from 30 to 50 bushels per acre. Crop insurance claims number more than 50,000 per year and, of those paid, 39% are for damage by drought, 14% for 'excess moisture', and 10% for hail damage, as against 11% for insect damage, 10% frost, 6% wind, and 5% disease.

REFERENCES

ATMOSPHERIC WATER RESOURCES PROGRAM [1968], *Project Skywater: 1967 Annual Report: Vol. I, Summary*; U.S. Department of the Interior, Bureau of Reclamation (Washington, D.C.), 79 p.

BAILEY, F. JR. [1965], Can you afford to lose your crops?; *The Farm Quarterly*, **20** (1), 66–7, 136, 138, 140, and 142.

BARRY, R. G. and CHORLEY, R. J. [1968], *Atmosphere, Weather and Climate* (Methuen, London), 319 p.

BATTAN, L. J. [1965] *Cloud Physics and Cloud Seeding*; The Science Study Series Number 27 (Heinemann, London), 144 p.

CURRY, L. [1952], Climate and economic life: A new approach with examples from the United States; *Geographical Review*, **42**, 367–83.

CURRY, L. [1962], The climatic resources of intensive grassland farming: The Waikato, New Zealand; *Geographical Review*, **52**, 174–94.

GOULD, P. R. [1963], Man against his environment: A game theoretic framework; *Annals of the Association of American Geographers*, **53**, 200–17.

HENDERSON, H. J. R. [1963], Climatic factors and agricultural productivity; *University College of Wales, Aberystwyth, Memorandum No. 6*.

HEWES, L. [1965], Causes of wheat failure in the dry farming region, central Great Plains, 1939–1957; *Economic Geography*, **41**, 313–30.

HORE, P. N. [1964], Rainfall, rice yields and irrigation in West Bengal; *Geography*, **49**, 114–21.

MASON, B. J. [1962], *Clouds, Rain and Rainmaking*; (Cambridge), 145 p.

MAUNDER, W. J. [1966], Climatic variations and agricultural production in New Zealand; *New Zealand Geographer*, **22**, 55–69.

MAUNDER, W. J. [1968], Effect of significant climatic factors on agricultural production and income: A New Zealand example; *Monthly Weather Review*, **96**, 39–46.

SEWELL, W. R. D., Editor [1966], Human dimensions of weather modification; *University of Chicago, Department of Geography, Research Paper 105*, 423 p.

SEWELL, W. R. D., KATES, R. W., and PHILLIPS, L. E. [1968], Human response to weather and climate; *Geographical Review*, **58**, 262–80.

SUGG, A. L. [1967], Economic aspects of hurricanes; *Monthly Weather Review*, **95**, 143–6.

TAYLOR, J. A. [1965], Weather hazards in agriculture; *The Royal Welsh, Annual Edition*.

VISHER, S. S. [1940], Weather influences on crop yields; *Economic Geography*, **16**, 437–43.

4.III(i). Water and Crops

ROSEMARY J. MORE

Formerly of the Department of Civil Engineering, Imperial College, London University

1. Plants and Water

Water is needed by plants to supply carbon dioxide and oxygen in solution to the cells for photosynthesis and respiration; to transport raw materials, manufactured and waste products within the plant; and to maintain the rigidity (*turgidity*) of the plant structure. Plant cells consist of a permeable *cell wall*, enclosing a layer of *cytoplasm* (a viscous fluid containing fine particles and

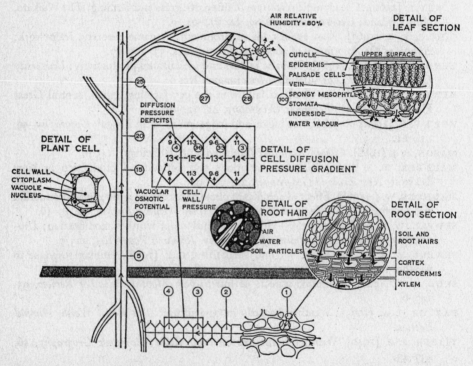

Fig. 4.III(i).1 Diagram illustrating the water circulation between soil, plant, and atmosphere. The possible diffusion pressure deficits at various points of this hydrostatic system are shown, assuming a 400-ft tree, readily available soil water, and an atmospheric relative humidity of 80% (Partly from Knight, 1965).

globules), which has the property of being *semi-permeable* (i.e. allowing the passage of some molecules but prohibiting others). The cytoplasm surrounds the *vacuolar sap*, an aqueous solution of salts, sugars, and organic acids (fig. 4.III(i).1). The more saline the sap, the lower the kinetic energy of the random movements of the water molecules, such that the solution is said to have a *diffusion pressure deficit* (D.P.D.) with respect to pure water. Therefore, where a permeable membrane separates two solutions of different salinity there is a net tendency for water molecules to pass from the less towards the more saline solution. The kinetic energy of the dissolved salt molecules, however, operates in the reverse sense, being greatest in the more saline solution. In flow between plant cells the semi-permeable property of the cytoplasm prevents the reverse flow of salt molecules, and only the net flow of water occurs, by a process called *osmosis*. The maximum force with which water can enter a cell is equal to the D.P.D. of the vacuolar sap, and were it not for the pressure of the stretched cell wall (W.P.), this would be identical to the *osmotic potential* (O.P.) of the sap. In reality:

$$D.P.D. = O.P. - W.P. \qquad (1)$$

Water moves in the plant as a continuous stream from the root hairs to the leaf surfaces in response to the cell *diffusion pressure-deficit gradient*. This gradient is made up of a chain of increasing D.P.D.s, and the D.P.D. is thus a measure of the capacity of a saline cell to absorb water, either from an adjacent (generally less saline) cell or from soil water. As equation (1) shows, the D.P.D. is not entirely a function of the cell osmotic potential, and water may pass between cells having equal osmotic potential (i.e. salinity), providing the cell-wall pressures are different. The cell-wall pressure is largely a function of the extent to which the cytoplasm is forced against the elastic cell wall by changes in the water content of the vacuolar sap. Cells can lose turgidity either by *plasmolysis* (see later) or by *wilting*. Wilting occurs whenever evaporation takes place from leaf or stem surfaces faster than water can be supplied to the cells, such that the vacuolar sap decreases, reducing the cell-wall pressure, causing the plant to droop.

The movement of water up from the roots to the leaves is controlled by its rate of evaporation (*transpiration*) from the leaf surfaces, and is thus termed the *transpiration stream*. One of the important functions of plant water is that the moist leaf surfaces absorb atmospheric carbon dioxide, but this maintenance of moisture implies a continuous water supply to the leaf surface to replenish that which is inevitably evaporated. Generally speaking, the upper surfaces of plant leaves are covered with a rather impermeable waxy *cuticle*, through which only about 10% of the total leaf transpiration occurs. The majority takes place through the cells of the small pores (*stomata*) in the *epidermis* on the leaf undersides. The rate of transpiration (assuming complete saturation of the leaf *mesophyll* spaces) depends, firstly, on the D.P.D. between the stomatal cell moisture and that of the air adjacent to the leaf (this being a function of the relative humidity of the air, the temperature of the air and leaf, and the rate of air movement), and secondly, on the size of the stomatal aperture (which differs between plants and is also controlled by the light intensity, such that stomata are closed and transpiration ceases at night). As the stomatal cell walls lose transpired moisture cell salinity increases, together with the osmotic potential of the cell, and shrinkage decreases the cell-wall pressure, both having the combined effect of increasing the D.P.D. and allowing available water to move by osmosis from adjacent cells. This process continues, in a highly complex manner, down the plant, and the water molecules from the weakly-saline soil water diffuse

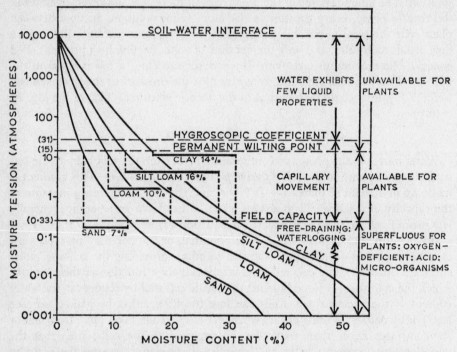

Fig. 4.III(i).2 The relationship between moisture content and moisture tension in four different types of soil, indicating the range of moisture available for plants (Partly from Buckman and Brady, 1960).

through the root cells into the *xylem*, the principal water-carrying cells of the plant.

Figure 4.III(i).1 shows this transpiration stream for a tall tree, the diffusion pressure-deficit gradient in a block of four root cells, and a series of cell diffusion pressure deficits at a number of points, indicating its gradual increase upwards, with the highest value occurring at the critical stomatal plant/air interface.

Returning to the link between the plant roots and the soil moisture, it is important to note that not all soil water is equally available for plant use. Figure 4.III(i).2 shows the moisture content (as a percentage of the total volume of the soil plus water) of four main types of soils, together with the soil-moisture tensions associated with each, indicating that as the interstitial soil-moisture films diminish, their tension increases, and it is less easy for the plant roots to extract moisture from the soil. After a rainstorm some of the water filling the soil interstices is free to drain away under gravity, leaving films held by a tension of about $\frac{1}{3}$ atmosphere. In this condition the soil is at *field capacity*, and plants are free to absorb both water and air from the soil. Unless the soil water is re-plenished, plant water use and direct evaporation from the soil combine to reduce the moisture content by shrinking the films. The accompanying increase of soil-moisture tension progressively denies the plant a freely-available water supply such that they may wilt in the daytime but regain turgidity at night (when the transpiration loss ceases). When the films have shrunk such that the tension of their surfaces reaches 15 atmospheres plants can no longer draw water from the soil, *permanent wilting point* has been reached, and the plant will ultimately die if water is not made available soon.

It is obvious that soil moisture exists in several degrees of utility for plants; that in excess of field capacity is superfluous, leads to oxygen deficiency, and needs to be drained; whereas, at the other extreme, moisture below the per-manent wilting point is not available for plant growth – indeed, at a tension greater than the *hydroscopic coefficient* (31 atmospheres) it loses most of its liquid properties. The amount of available soil moisture depends partly on the charac-teristics of the plant and partly on the soil type. Figure 4.III(i).2 shows the percentage of *available water* in four main soil types, sand having the least (7%) and silt loam the most (16%). Finer-textured soils can hold more available moisture because of a combination of greater total pore space, the greater wet-table surface area of the soil particles and, usually, a greater proportion of water-retaining colloidal matter. Factors affecting the ability of plants to avail them-selves of soil water are the extension of the root system, the drought-resistant properties of the plant, and the stage and rate of its growth. Plant roots only draw soil water from their immediate neighbourhood. It is not clear whether soil moisture is equally available to a plant throughout the whole range from permanent wilting point to field capacity, and it has been suggested that the *optimum moisture zone* occurs at a moisture content considerably above the permanent wilting point.

2. Plants as crops

Since the Neolithic Revolution man has cultivated many plants as crops. Of the four factors significant in cropping, the plant, the atmosphere, the soil, and its water, man cannot modify the atmosphere (except in the most limited sense), and he can do only a small amount to modify the physical and chemical composition of the soil over large areas (e.g. by the application of fertilizers). However, he can exploit and utilize the genetic variations within plant species and selectively breed those varieties of plants which are resistant to moisture and temperature extremes, deficiencies in soil chemistry, the activity of pests, etc. Some of the most striking changes in cropping are made, however, by the degree to which man can control the amount and quality of water available to plants, either by irrigation, drainage, or improvement of water quality. His aim is to keep soil moisture as close as possible to the optimum for the soil type and the crop, within the limits of engineering and economic feasibility. However, in extensive semi-arid areas of the world man has found it uneconomical to make appreciable modifications to the soil-water environment and has developed *dry farming* techniques in association with drought-resistant plant strains. Cultivation loosens the soil and allows it to absorb and retain more water than if it remained in a dense untilled state. In the Great Plains of the United States, for example, dry farming involves in fallow year deep early-spring tilling which allows greater absorption of the succeeding rainfall and limits surface evaporation losses. The accompanying increase in weed growth is counteracted by periodic cultivation or the use of weed-killers. This exploitation of the cultivated soil as a moisture-storage reservoir during a fallow year conserves a variable amount of the annual rainfall, usually about a quarter, and after such a fallow year wheat yields in Kansas and North Dakota, for example, may be almost doubled. Dry farming, however, is still at the mercy of annual rainfall variations, and the amount and quality of yields inevitably strongly reflect this variability.

3. Irrigation

The purpose of irrigation is the control of soil moisture between a lower limit that will not restrict plant growth and an upper limit that avoids the disadvantages of water-logging. One calculation of irrigation need is based on the concept of *potential transpiration*. Penman has shown that during the May-to-September growing season in Western Europe the energy used in transpiration is the largest single term in the net exchange of energy between the sky, atmosphere, and ground. The following seven terms are involved in the energy balance (H) (the percentages giving the relative orders of magnitude recorded in a series of British observations during 1949):

Total incoming solar radiation (R_c) (100%)
Transpiration (E) (39%)
Back radiation from the surface (R_B) (34%)
Reflection from the surface ($R_c . r$: where r = the albedo) (20%)
Heating the air (K) (4%)
Heating the soil (2%)
Plant growth (1%)

Of these, the last two are small enough to be omitted, leaving the following relationships:

$$H = E + K \simeq R_c - R_c . r - R_B \tag{2}$$

in which R_c, r, and R_B can be determined from meteorological observations (duration of bright sunshine, air temperature, humidity, and cloudiness) and K can be eliminated by expressing it as a ratio of E. The ratio of E_0/K (where E_0 is the estimated evaporation from a theoretical open water surface) is obtained by inserting measured meteorological values (wind-speed variations; water temperature and saturation vapour pressure; and air temperature and vapour pressure) into other equations. E_0 can be transferred to the *potential transpiration* (E_T, which is considered equal to E) by recognizing that the rate of transpiration from a short green crop (e.g. grass) with an adequate water supply is the same as the evaporation rate from a water surface, and by merely supplying a weighting factor (for Britain during May to September $E_T/E_0 = 0.8$) related to the length of daylight during which the stomata are open and transpiration takes place. Substitution into equal (2) thus gives the amount of potential transpiration which would occur with an adequate water supply, but when this is not available any deficiency of precipitation with respect to potential transpiration forms a *moisture deficit* which, for adequate cropping, must be made up from pre-existing soil-moisture supplies or from applied irrigation water.

There are two restrictions to the wide application of this ingenious approach to the calculation of plant water needs, firstly, the lack of detailed meteorological measurements in many localities and, secondly, that it is less well adapted to regions of high temperature, low humidity, and sparse vegetation and crop covers. In the western United States a simplified calculation for crop water use (shown as *consumptive use*) has been devised by Blaney and Criddle, such that:

$$u = kf \qquad (3)$$

where u = monthly consumptive use (in inches);

k = consumptive use coefficient dependent upon mean monthly temperature and crop stage of growth;

f = monthly consumptive-use factor $\dfrac{t \times p}{100}$;

t = mean temperature in °F;

p = percentage of daytime hours of the year, occurring during the period (from tables).

The following Table 4.III(i).1 gives the values used in calculating monthly consumptive use by alfalfa in the Upper Salinas Valley of California:

TABLE 4.III(i).1

Month	t	p	f	k	u
April	57·9	8·85	5·12	0·60	3·07
May	62·5	9·82	6·14	0·70	4·30
June	65·7	9·84	6·46	0·80	5·17
July	68·4	10·00	6·84	0·85	5·81
August	67·8	9·41	6·38	0·85	5·42
September	66·6	8·36	5·57	0·85	4·73
October	62·2	7·84	4·88	0·70	3·42
Total consumptive use for the irrigation season					31·92

Irrigation is used to supply any deficit existing between potential transpiration and precipitation. The engineering techniques by which this is accomplished vary according to the type of crop and the physical environment, but are mostly controlled by the economic level of the society concerned. Primitive irrigation usually involves local, temporary storage or diversion of surface runoff, whereas in advanced societies irrigation is economically linked to multiple-purpose

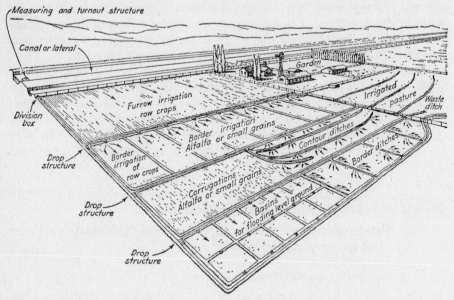

Fig. 4.III(i).3 Irrigation techniques. Various methods of applying irrigation water to field crops (From Israelson and Hansen, 1962).

projects concerned with large dams, canals, and a variety of sophisticated water-spreading techniques (fig. 4.III(i).3).

4. Drainage

The roots of most plants and cultivated crops develop above the perched soil water-table, where both air and water are available. However, in some fine-grained soils, particularly clays, the available air in the interstices below the capillary fringe is so meagre that plant roots cannot penetrate much below this fringe, and there is evidence that most crop roots do not extend deeper than some 12 in. above the average height of the water-table in the soil. Saturation results in a reduction in the absorption of soil oxygen and plant nutrients, leading to inefficient photosynthesis and consequent reduction in transpiration. Artificial drainage is thus aimed at lowering the soil water-table to allow a sufficient depth of the aeration zone wherein a healthy root system can develop well supplied by water, air, and soil nutriments. The optimum water-table and rooting depths differ between crops, the former being about 20 in. for celery and 36 in. for cereals and sugar beet in fen peat soils of Eastern England, but fig. 4.III(i).4 provides some general relationship between crop yield and depth of water-table for four soil types in Holland, showing that as the soil texture becomes heavier maximum yields are obtained with deeper water-tables. Some crops are very water tolerant, for example, rice, which has a mechanism for obtaining oxygen direct from the air, but others, like tobacco, are very sensitive to excess water. Soil drainage also has other generally beneficial effects on crops by increasing the nitrogen supply which can be obtained from the soil and by changing the thermal properties of the soil. Drainage decreases both the specific heat and the thermal conductivity of soils, together with the amount of solar energy lost in

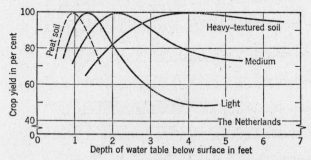

Fig. 4.III(i).4 Depth of water versus crop yield. The general relationship between crop yield and constant water-table depth during the growing season in the Netherlands (After Visser: From Schwab, G. O. *et al.*, 1966, *Soil and Water Conservation Engineering* (Wiley, New York)).

evaporation from soils, and has the net effect of making more soil heat energy available for plant growth.

From the foregoing it is apparent that it is usually more difficult to calculate optimum drainage requirements than optimum soil moisture or water-quality requirements.

5. Improvements in soil water quality

Quality of soil water is as important as quantity in achieving optimum plant growth, for certain amounts of dissolved salts derived from mineral and organic soil constituents (and fertilizers) are vital for plants. In humid soils the ready availability of water generally causes the soil water to be dilute, but in arid regions short intense storms do not permit infiltration into more than 1–2 ft of the soil, and the rapid direct evaporation of the rainfall leads to higher salt concentrations in the upper part of the soil profile which exert injuriously high osmotic potentials. The accompanying shrinkage of water films near the surface causes saline water to be drawn up from depth (as long as the continuous film stage exists), the evaporation of which leaves the soil and soil water more saline. Salinity can also be inadvertently increased artificially by applying saline irrigation water. Salinity is controlled by the amount of soluble salts (chiefly sodium, calcium, magnesium, and potassium) which are present, and can be measured

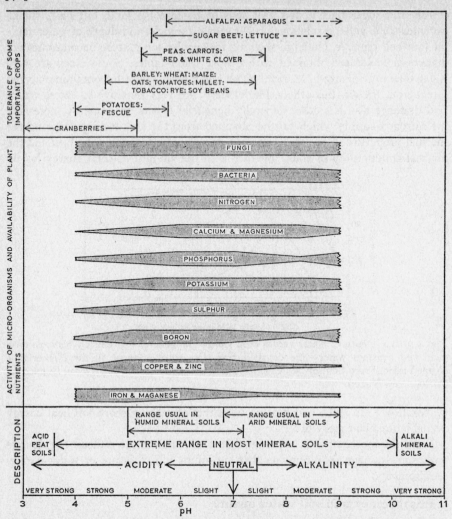

Fig. 4.III(i).5 The relationship between the activity of micro-organisms and the availability of plant nutrients, on the one hand, and of mineral soil pH. The width of the band represents the pH ranges of greatest activity and availability. The pH tolerances of some important crops are also shown (Adapted from Buckman and Brady, 1960).

in the field by simple electrical instruments because salinity exerts a direct control over the electrical conductivity of water.

If plant roots are in contact with soil water of higher salinity than that of their vacuolar sap osmotic withdrawal of the sap causes the cytoplasm to shrink away from the cell wall and the space between to be invaded by saline soil water (which passes into the cell through the permeable cell wall). This process of *plasmolysis* can be reversed and turgidity restored if soil water of sufficiently low salinity to allow the cell to absorb water molecules by osmosis becomes available.

TABLE 4.III(i).2

Source	Water class (U.S. Salinity Lab.)	Total Salt content		Electrical conductivity (micromhos per cm)	Some important constituents			Comments
		(p.p.m.)	(Tons/acre-ft)		Calcium and magnesium (m.e. litre)	Sodium (%)	Boron (p.p.m.)	
Pecos River, Texas	3	6,198	8·4	9,150	47·8	52	—	Harmful to most crops. Here counteracted by natural soil lime and gypsum, fair crops of cotton and alfalfa
Coachella Valley, California, ground water	2 Plus excess sodium and boron	910	1·2	1,740	2·2	85	0·71	Excess sodium and boron injurious to some crops (e.g. beans and grapes)
Rio Grande, New Mexico	1	641	0·8	870	5·1	4·0	—	Suitable for most plants

The availability to the plant of chemical and biological nutrients from the soil, besides being obviously related to soil mineral composition, temperature, etc., is largely influenced by the pH values of the soil water (fig. 4.III(i).5). Just as there is an optimum range of soil water, so the pH range of 6–7 is optimum for plant growth, below which important nutrients (notably phosphorus) are unavailable in the acid soils, and above which occur the alkaline soils of the arid and semi-arid regions.

Such arid soils, which present special water-quality problems, can be divided into:

(a) Saline soils: with excess soluble salts, producing plasmolysis and impairing productivity.

(b) Saline/alkali soils: with excess soluble salts plus an injurious excess of sodium.

(c) Alkali soils: with an excess of sodium and pH of 8·5–10.

Treatment of such soils is largely effected through a control of soil water by:

(a) Lowering the soil water-table so that soluble salts can be leached away by rainfall or applied irrigation water. This is the standard treatment for saline soils, but with saline/alkali and alkali soils a soluble calcium compound (e.g. gypsum) must first be added to leach away the sodium salts to improve the permeability so that leaching can proceed.

(b) Not applying irrigation water to excess, thus keeping the water-table low and avoiding drawing up salts to the capillary fringe near the surface.

(c) Cultivation designed to promote free percolation of surface water into the soil and preventing waterlogging.

(d) Control over the quality of applied irrigation water.

The quality of irrigation water in arid regions is commonly alkali, and it is necessary to apply it in controlled amounts to crops with appropriate levels of salt tolerance (fig. 4.III(i).5). The table 4.III(i).2 exemplifies three classes of irrigation water. Class 2 and 3 water must be applied with care to crops, because the salt concentration of soil water is commonly 2–100 times that of the applied irrigation water (depending on the amount of direct evaporation from the soil), and the accumulation of salts may build up, ultimately making the soil unsuitable for cropping, as in large areas of West Pakistan.

6. Interactions of irrigation, drainage, and water quality

It is obvious that these three aspects of water management are interdependent, and that a balance must be achieved to provide an optimum moisture environment for crops, both as regards quantity and quality. A classic example of integrated water management is the recent reclamation of an area in West Pakistan, made saline by previous uncontrolled irrigation. This was effected by a balance of water-table reduction by pumping, combined with the application of just enough irrigation water to allow leaching without waterlogging.

Water management is only one aspect of total crop management, however, and must be combined with fertilizer application, pest control, good soil management, selection of appropriate plant strains, and other sound agricultural practices in an effort to achieve maximum productivity within the limits of the plant's total environment.

REFERENCES

BUCKMAN, H. O. and BRADY, N. C. [1960], *The Nature and Properties of Soils* (The Macmillan Co., New York), 567 p.

FOGG, G. E. [1963], A digression on water economy; Chapter 3 in *The Growth of Plants* (Penguin Books, London), pp. 61–87.

KNIGHT, R. O. [1965], *The Plant in Relation to Water* (Heinemann, London), 147 p.

ISRAELSON, O. W. and HANSEN, V. E. [1962], *Irrigation Principles and Practices* (John Wiley & Sons, New York), 447 p.

LUTHIN, J. N., Editor [1957], *Drainage of Agricultural Lands* (Madison, Wisconsin), 620 p. (especially Chapter 5).

MORE, R. J. [1965], Hydrological models and geography; In Chorley, R. J. and Haggett, P., Editors, *Models in Geography* (Methuen & Co., London), pp. 145–85.

PENMAN, H. L. [1963], *Vegetation and Hydrology* (Commonwealth Bureau of Soils, Harpenden), Technical Communication No. 53, 124 p.

4.III(ii). Primitive Irrigation

ANNE V. KIRKBY

Department of Geography, Bristol University

Primitive irrigation is characterized by its great variety; in many different physical environments it is practised by peoples with varying cultures and levels of technology and on scales ranging from individual farmers to the great hydraulic societies of India and China. It includes the flooded ricefields of the Tonkin delta, the small maize patches of the Hopi Indians in Arizona, the water meadows of medieval England, and the wheatfields of ancient Persia. Despite the variety of physical settings and methods used, most primitive irrigation schemes show a sophisticated awareness of local hydrological conditions and are economic in that they produce an increased return over rainfall farming for the increased input of labour and equipment. Once a community has decided to irrigate, the irrigation system employed is chosen within two sets of constraints – those of the physical environment and those of their own culture, especially the level of technology and the economic framework. Conversely, once the method of irrigation is decided in accordance with these constraints, the irrigation system itself can set constraints on the economy and society of the irrigators and can locally modify the physical environment so that an important feedback between water and man takes place through irrigation. In its ultimate form, absolute state power may be achieved through control of water resources and their distribution.

Man may move water in two ways: in units, such as bucketfuls, or as a continuous flow, as in a canal. The movement of water for irrigation may be divided into three phases: (1) abstraction from the water source; (2) distribution throughout the irrigation system; and (3) application on to the crops. In each of these three phases the water may be moved as units or as a continuous flow, and many primitive irrigators combine both types of movement within one system. The sources of water available for irrigation are limited by the physical environment, the most common natural sources being rivers, lakes, springs, and ground water, but where these are not available primitive irrigators may depend on more unusual sources; for example, water stored in limestone sink holes or in sand dunes. In primitive systems water from surface sources is generally abstracted by means of continuous flow, in canals, and water from ground-water sources, which requires lifting to the field surface, is most easily taken out in units. However, within the great range of primitive irrigation schemes known, examples may be found of lifting surface water in units, as is done from rivers in India, by means of a lever device called the *shaduf*; and ground water can be brought to the surface as a continuous flow without the use of modern pumps, by constructing a complex system of wells and tunnels through which the ground water flows to the surface under gravity, as in the *ghanats* of Iran and the *galerias* of Tehuacán, Mexico (fig. 4.III(ii).1).

The second, or distributional, phase of the irrigation scheme is usually done by means of canals, which vary in construction and efficiency from unlined,

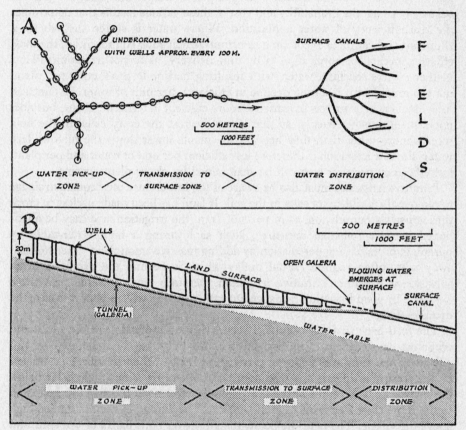

Fig. 4.III(ii).1 Typical *galeria* irrigation system from the Tehuacán Valley, Mexico; (A) plan, and (B) cross-section.

meandering earth channels with high evaporation and seepage losses, to tiled or cemented canals and aqueducts which minimize water losses. Water can be distributed without using canals, and in some schemes it is moved underground through tunnels (*ghanats* and *galerias*), and in pot irrigation in Mexico it is distributed in pots or buckets which the irrigator carries from the nearest of several wells in his small plot to each plant individually.

The third phase of the system, that of applying the water to each plant, employs one of the following methods:

1. individual canals (including furrows);
2. standing sheets of water (e.g. swamp rice cultivation);
3. flowing sheets of water (e.g. simple flood water farming);
4. unit application by pots or buckets.

Each of these methods have different efficiencies in terms of crop production per unit of irrigation water, per man hour, per plant, and per unit area, and it is the balance between the availability and cost of these various factors that determines the real efficiency of water application. Where water is scarce and labour is abundant and cheap, as is common in primitive irrigating communities, the most efficient method of application is by unit delivery, as in pot irrigation. Plants neither receive too little water, with resulting decline in production per plant, nor too much, with resulting decline in efficiency per unit of water. In contrast, individual canal or furrow irrigation is more efficient in terms of labour, but does not distribute water evenly, so that plants nearest the entry canal in the field receive more water than they need, while plants lower down the furrows lack water. Furrow irrigation is therefore less efficient per unit of water and per plant, but efficiency per unit area can be maintained by closer plant spacing.

Primitive irrigators must also be aware of the importance of adequate drainage to remove the build-up of salts in the soil. If land has been made useless to crops through salt accumulation, as in parts of Iran, the irrigation area may be abandoned because remedial measures, such as lowering a high water-table by pumping, or increasing percolation by adding massive amounts of water to leach away salts, often require capital outlay which is beyond the primitive near-subsistence irrigator. Primitive irrigation therefore tends towards preventive medicine by good irrigation management rather than drastic measures to save the dying patient.

The feed-back that *can* occur between irrigation methods and the social and economic development of the irrigating community may be illustrated by two villages in the semi-arid Valley of Oaxaca in Mexico. The first village is situated on the crest of a ridge in the piedmont zone, where the water-table is far below the surface but where one of the few perennial streams in the area can be led on to the ridge crest by means of a take-off canal beginning in the mountains high above the village. The fields lie below the village on both flanks of the spur and are irrigated by a distributary system of earth canals and finally by means of individual furrows. The take-off and main canals are communally owned and maintained by the farmers of two villages, and because the water is distributed from a single source which cannot irrigate all the fields at the same time it is allocated to farmers on a rota basis of once every fifteen days. Crops are therefore selected from those which grow best with fifteen days between waterings so that there is a strong tendency for all farmers to grow the same irrigated crop, alfalfa, and the resulting land-use pattern is very uniform. The system of water distribution not only establishes the necessity for, and the idea of, co-operation between farmers but also establishes the machinery, in the form of water overseer and committee, for organizing community action in other spheres and for enforcing the majority opinion on the minority by the threat of withdrawal of water rights. Community projects, such as road building and the installation of electricity, therefore play an important role in the life of this village.

The second village, although only eighteen miles away, is in strong contrast to the first. It is located on the flat valley floor away from the mountains, where there is no perennial surface water, but a high water-table at between 5 and 12 ft deep provides a widespread source of irrigation water. Instead of receiving water from a single source via a communal distributory system, farmers can pot irrigate from many shallow wells. Thus abstraction, distribution, and application are all within each farmer's individual control, and he requires little co-operation with his neighbours for the success of his irrigation scheme. The amount and frequency of watering may be adjusted to the requirements of each crop and even to each plant, and is controlled by each farmer so that the range of crops grown is diverse and the resulting land-use pattern is very varied, in contrast to the monotony of alfalfa fields in the piedmont village. There is little co-operative enterprise, and no machinery exists for enforcing co-operation, so that community action plays a small role, and even schemes for the direct benefit of farmers, such as flood-water control and the co-operative ownership of irrigation pumps, fail in the face of the aggressive independence of family groups.

In these two Mexican villages therefore two very different irrigation schemes have resulted from, on the one hand, a single water source from which water must be distributed to the fields; and on the other hand, a widespread water source which is readily obtainable within each field. In such primitive irrigation communities as these there is therefore a real two-directional interrelationship between water and man, between the physical environment and the economic and social life of the communities which exploit it.

Acknowledgement. To Professor Aubrey Williams, Dept. of Sociology and Anthropology, University of Maryland, U.S.A., for ethnographic information about the Valley of Oaxaca, Mexico.

REFERENCES

CLARK, C. [1967], *The Economics of Irrigation* (Oxford), 116 p.

ISRAELSEN, O. W. and HANSEN, V. E. [1962], *Irrigation Principles and Practices;* 3rd Edition (New York), 447 p.

KIRKBY, A. V. [IN PREPARATION], *Land and Water Use in Present and Ancient Oaxaca, Mexico;* unpubl. Ph.D. thesis, The Johns Hopkins Univ., U.S.A.

WITTFOGEL, K. A. [1957], *Oriental Despotism* (New Haven, U.S.A.), 556 p.

5.III. Overland Flow and Man

M. A. MORGAN

Department of Geography, Bristol University

It is obviously impracticable to draw too fine a distinction between the movement of water over the surface and immediately beneath the surface, a distinction, that is, between overland flow and throughflow, since in practice the one shades imperceptibly into the other. The infiltration capacity of a soil, and hence the relationship between overland flow and throughflow, depends on the texture of the soil, on its thickness and its degree of compaction. Surface runoff can only occur when the intensity of the rainfall exceeds the infiltration capacity of the soil. A clay soil composed of many fine particles and containing much interstitial water has a low infiltration capacity, while a dry sandy soil will at least for a time absorb and pass water fairly easily (Table 5.III.1).

TABLE 5.III.1 Infiltration capacities of field soils

Soil texture	Infiltration capacity (in./hr)
Clay loam	0·1–0·2
Silt loam	0·3–0·6
Loam	0·5–1·0
Loamy sand	1–2

(after Kohnke and Bertrand)

Hard driving rain will cause compaction of the surface of a normally permeable soil, reducing the pore spaces and thereby lowering the infiltration capacity. A good vegetation cover will prevent compaction by rain, and thus increase the infiltration capacity. Obviously climate, geology, soil, and vegetation all affect the character of runoff.

The ideal situation from the point of view of man is where ample gentle rain falls at the right time for his crops, penetrates the soil without violence, maintaining just sufficient moisture around the soil particles without filling the interstices, and passes down to nourish a stable water-table, which in turn feeds a system of streams and rivers whose regime is accommodating and predictable. Unfortunately even in the most favoured areas these conditions are rarely completely satisfied. The runoff in a sense may be regarded as a measure of the extent to which any particular hydrological system departs from the ideal from a human point of view. Too great, irregular, or unpredictable runoff shows that the system is not working at its best for the farmer. A great deal of the history of farming is the story of the attempts to find ways of improving the plant–soil–water relationships, and most of the world's productive agricultural areas show the imprint of generations of effort directed to redressing some of the imbalances caused by undesirable runoff characteristics.

1. Soil erosion

Certainly in human terms the most far-reaching and devastating consequence of excessive uncontrolled runoff is soil erosion, and there are many areas in which constant vigilance against its effects are essential. In the humid and sub-humid regions natural vegetation usually binds the soil particles in place and slows down the flow of surface water. Although erosion takes place, natural processes normally create at least as much soil as they remove. Interference by man with the natural vegetation, which is usually finely and uniquely adapted to the totality of natural conditions, can start the process of erosion. Once bare soil is exposed on sloping land, whether by ploughing, overgrazing, or by too enthusiastic felling of timber, it is vulnerable. Wherever rain falls faster than it can soak into the bare earth the raindrops loosen the soil particles. The effectiveness of raindrops in detaching soil particles depends on the intensity of the rainstorms and at what season they occur. Generally summer thunderstorms are the most violent. High rainfall intensities produce relatively large raindrops, which have high velocities, and these in turn can detach more and bigger fragments of soil. Such particles are borne away by the surface water as it flows downslope, carrying with it mineral salts in solution as well as organic material. The wholesale removal of layers of soil from slopes is known as sheet erosion. Where the process is maintained for any length of time the water is gradually concentrated into definite channels, giving rise to rill erosion. All too frequently in the past, in the absence of remedial measures the rills in turn have grown into gullies, which can extend at rates of a hundred feet a year, carving deep into farmland and often making it impossible to use machinery. Streams and rivers become choked with sediment, which they spread in a suffocating blanket over cropland in the valley floors. The loss of texture and porosity in the denuded soils that are left on the slopes makes them even less capable of absorbing the rain, so the runoff increases with time, and the regimes of the streams and rivers become even more erratic. The nature of the soil, the severity of the slopes, the geological structure, the rainfall characteristics, and the type of farm enterprise all play their part in affecting the nature and extent of soil erosion in any one area. Table 5.III.2 represents an attempt to indicate in very simple terms the relationships between the more important variables, and is formulated in terms of the conditions in the United States. While it is easy to overdramatize the effects, it is none the less the case that in the United States alone in the late 1950s erosion was estimated to be the main or dominant conservation problem on over 700 million acres, more than half the agricultural land in the country, and 234 million acres of cropland needed constant attention against erosion.

TABLE 5.III.2 Classification of runoff-producing characteristics

Designation of watershed characteristics	Runoff-producing characteristics			
	100 Extreme	75 High	50 Normal	25 Low
Relief	(40) Steep, rugged terrain; average slopes generally above 30%	(30) Hilly; average slopes of 10-30%	(20) Rolling; average slopes of 5-10%	(10) Relatively flat land; average slopes of 5%
Soil infiltration	(20) No effective soil cover; either rock or thin soil mantle of negligible infiltration capacity	(15) Slow to take up water; clay or other soil of low infiltration capacity, such as heavy gumbo	(10) Normal, deep loam; infiltration about equal to that of typical prairie soil	(5) High, deep sand or other soil that takes up water readily and rapidly
Vegetal cover	(20) No effective plant cover; bare or very sparse cover	(15) Poor to fair; clean-cultivated crops or poor natural cover; less than 10% of drainage area under good cover	(10) Fair to good; about 50% of drainage area in good grassland, woodland, or equivalent cover; not more than 50% of area in clean-cultivated crops	(5) Good to excellent; about 90% of drainage area in good grassland, woodland, or equivalent cover
Surface storage	(20) Negligible; surface depressions few and shallow; drainage ways steep and small; no ponds or marshes	(15) Low, well-defined system of small drainage ways; no ponds or marshes	(10) Normal; considerable surface-depression storage; drainage system similar to that of typical prairie lands; lakes, ponds, and marshes less than 2% of drainage area	(5) High; surface-depression storage high; drainage system not sharply defined; large flood-plain storage or a large number of lakes, ponds, or marshes

Each column shows the contribution of the different watershed characteristics, relief, soil infiltration, vegetation, and surface storage to a particular amount of runoff. For example, under extreme conditions of 100% runoff, relief accounts for 40% of the runoff, while the other three elements account for 20% each. The right-hand column shows that a low proportion of runoff is considered to be 25%, to which relief contributes twice as much as either soil infiltration, vegetation, or surface storage.

From *Farm Planners' Engineering Handbook for the Upper Mississippi Watershed*, U.S. Soil Conservation Service, Milwaukee, Wis., 1953.

Nor is erosion by running water the only hazard. The exposure of bare earth to strong drying winds can result in the soil particles being detached and carried along by the wind. Very small particles less than 0·1 mm in diameter can be knocked off the surface by larger ones in motion, and carried off in suspension distances of several hundreds, even thousands, of miles. Sand-sized particles varying in diameter between 0·1 and 0·5 mm are rolled along by the wind. Since there is no wind at all at the actual surface of the ground, only the tops of such particles are directly affected by the wind, and they start to spin, reaching several hundreds of revolutions a second. As a particle rolls a partial vacuum is created above it, lifting it into the air rather like a plane when it reaches take-off speed. As it rises the particle loses lift and falls back to the ground. In fact, it moves with the wind in a series of bouncing hops, a process known as saltation. Soil fragments slightly larger, between 0·5 and 3·0 mm in diameter, are too heavy to be lifted, but they can be rolled along the ground, causing surface creep. Only exceptional winds can move particles bigger than 3 mm in diameter. Sandy soils, whose particles are mostly the best size for saltation to be effective, are most liable to wind erosion when conditions are right, but since wind erosion is selective of particle size, the finest and often the most fertile parts of good soils, the smallest fragments, can be progressively winnowed out by the wind, leaving behind an impoverished and coarse soil whose fertility is much reduced. The Dust Bowl is the classic area in the United States, and was created initially by too regular cropping for wheat and by the practice of leaving the soil exposed during the period of summer fallow.

Normal conservation practices, largely developed in the United States, are based on watersheds as the management units. They are devoted to stopping active erosion and to reducing runoff by increasing the infiltration capacity of the soil. Conservation methods include soil treatment, introduction of improved crop rotations, contouring and terracing, gully damming, drainage diversion, grassing waterways, cover cropping, pasture improvement, tree planting, and woodland conservation. Conservation schemes are costly, but most governments have systems of subsidies and grants to encourage farmers to make improvements. The economic and social benefits of such conservation programmes are often direct and immediately apparent, and this tends to decrease the hostility with which rural communities might normally be expected to regard innovations. Many schemes pay for themselves in a few years. For example, in many parts of the United States it is normal practice to dam streams in the upper reaches of drainage basins. The lakes thus formed are used to store runoff water. But they also serve other purposes as well. They can be used for recreation, and by furnishing emergency water supplies for livestock and for fighting farm fires they can directly benefit the farmer by lowering his costs. They also reduce the cost to the Government of flood compensation. Even so, it must be admitted that social and institutional resistance to improvement schemes of this kind are still a force to be reckoned with in the United States, let alone in other countries, whose rural communities are often far less sensitive to economic and financial incentives.

While attention has properly been focused on the problems of soil erosion in rural areas, it is in some ways almost as great a problem in some of the rapidly expanding urban and suburban areas. Construction sites for motorways, freeways, factories, and housing estates may have acres of raw earth stripped and exposed by bulky excavators and earth-moving equipment. Such sites can remain open for several years, even the single dwelling creates its own small problem for several months at least. Some cities in the United States have regulations designed to reduce erosion rates under such conditions, but they are usually neither comprehensive nor easy to enforce. Official practice in conservation-conscious areas encourages temporary cover of exposed places, with annual sown grasses or even tarpaulin, and drainage ways to take storm water gradually off the site by way of sediment traps, so that as little top-soil as possible is lost. Reliable information on rates of loss is not easy to find, but a recent study by Wolman and Schick of conditions on construction sites in Maryland, where soils are deep, slopes generally less than 10%, and annual rainfall about 42 in., suggests that rates are between twice and several hundred times as great per unit area as from the surrounding rural areas, and 'the equivalent of many decades of natural or even agricultural erosion may take place in a single year from areas cleared for construction'. Comparative figures suggest a typical sediment yield from undisturbed areas of between 200 and 500 tons per square mile per year, and on construction sites figures range from several thousand up to 140,000 tons per square mile per year. The very high yields are recorded on very small sites, and in general the bigger the site, the smaller the sediment yield, since the big site allows internal readjustment of material, much of which never gets carried beyond the site to contribute to the load of external rivers and streams. The increased stream loads give rise to many problems. Domestic and industrial water supplies are often drawn from this source, and where the water is polluted by increased turbidity additional costs are incurred in removing contamination. Reservoirs silt up more rapidly. The ecology of rivers and ponds can be changed very quickly as the channels accommodate themselves to the increased burden of sediment and fishing and sporting interests are naturally affected. Recreational land use possibly suffers most of all, and though it is almost impossible to put a value on amenity, it is beyond dispute that recreational land near the margins of the large cities grows more precious each year, and anything that contributes to despoiling such amenities is not only to be deplored but to be resisted.

2. Rural drainage

Successful farming in much of the temperate area in the world depends to an extent much greater than commonly realized upon the success of schemes for getting rid of excess soil moisture. Plants only use the water that forms a thin film around the soil particles. The interstitial water is a nuisance, its presence makes the soil heavy, and the object of any drainage scheme is to get rid of it, to lower the water-table, thereby raising the soil temperature through reducing evaporation, and at the same time aerating the soil.

Many of the oldest drainage schemes in Britain were undertaken by the early shepherds on the mountain and hillslopes. Over the course of many centuries they created a network of drainage channels designed to remove surplus water from the grazing areas in such a way that it did the least possible damage. On flattish land, however, too intricate a network of drainage ditches impedes cultivation, the ditches need frequent attention to clear them of aquatic plants, and by themselves they will not usually deal effectively with really wet soils. One of the greatest but often underestimated drawbacks of the medieval open-field system of farming was that it was not easy to drain the great fields effectively. The open ditches and deep furrows that the farmers made were all too often hopelessly inadequate, and after heavy rain water could stand on the land week after week, gnawing away at its fertility.

Another and usually superior way of draining superfluous moisture from the soil is by the use of underdrains. The Enclosure movement in Britain created compact individual land holdings in much of the area previously dominated by the open-field system, and released from the restrictions of a communal system, individuals were free within limits to experiment with new methods to drain the heavy lands. The first textbook in Britain devoted to the study of farming techniques was probably *The Improver Improved*, by Capt. Walter Bligh, published in 1650. It is apparent from this and later books that underdraining was already being practised in some areas. One popular method of the time consisted of cutting parallel trenches from 18 to 48 in. deep so arranged as to run down the slope. The bottoms of the trenches were filled with a layer of blackthorn, other brushwood, or straw, covered with a turf, which in turn was covered with stones and then earth until all traces of the trench had been removed. In time the replaced soil would consolidate above the brushwood or straw, so that when that eventually decayed a tunnel would be left below the surface through which surplus water would drain to a suitable outfall on the edge of the field. Stones were used instead of brushwood where they occurred locally in sufficient quantity. Arthur Young, in his travels in England in the latter half of the eighteenth century, noted with approval the extensive use made of drains of this type. Agricultural techniques at the time were still composed of a blend of traditional practices, on the one hand, and the enlightened empiricism of a few men of vision operating without the benefit of a truly scientific background, on the other, so we need not be surprised that many different approaches to the problems and purposes of underdrainage all found the active support of enthusiastic advocates, and nearly all had the merit that at least they worked. In 1831 James Smith of Deanstone in Perthshire published an article explaining the methods of drainage by means of which he succeeded in the space of a few years in transforming his marshy and sour farmland into a rich and fertile property. His work attracted the attention of a Select Committee of the House of Commons charged with an inquiry into the State of Agriculture at a time of agricultural distress in 1836. The report devoted much of its space to an investigation of drainage, and its chairman at one point ventured the (probably excessive) opinion that drainage was 'the only thing likely to promote the general improvement of agriculture'. The report is a valuable guide to the nature and extent of the drainage methods in use at the time.

Smith's system was designed both to drain and to break up the hard sub-soil pan that develops in many wet soils. Parallel field drains were laid 30 in. deep and between 16 and 21 ft apart. Then Smith used his own design of plough, a formidable implement drawn by a large team of horses to cut deep into the soil, crumbling the hard pan, breaking up the sub-soil, aerating it, warming it, and encouraging it to yield up its water more easily to the underlying drains. Josiah Parkes (1793–1871) believed that deeper drains were preferable, and there followed a period of controversy sustained for several years in the fashion of the

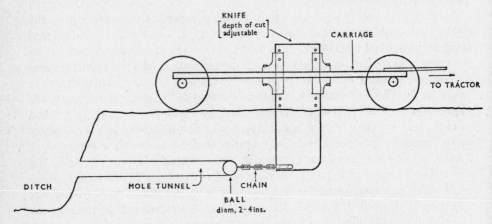

Fig. 5.III.1 Method of forming a mole drain.

times and nourished by a succession of passionately partisan lectures, papers, and articles.

An important advance was marked by the invention of a cheap, seamless clay pipe in 1843. On large estates with deposits of suitable clay it was common practice to build tile factories as near as possible to where the pipes were required. Cartage of pipes and 'tiles' was expensive, and where no local clays were found stone or straw continued to be used in trenches for several decades. Landlords found that their tenants were in many cases not averse to paying increased rents for drained lands. Increases of 5s. per acre with drains at 18-ft intervals and 2s. 6d. per acre with drains at 36-ft intervals were quoted in 1836 for tenants of the Duke of Portland in Ayrshire.

With the advent of steam engines many complicated and ingenious machines were invented to trench, lay the clay pipes, and fill in in a single operation. Where the soil was a tough homogeneous clay and where the slopes were fairly simple, it was easier and cheaper to use the mole plough. This was developed towards the end of the eighteenth century, and is still used (fig. 5.III.1). As it is drawn through the soil it leaves a 'pipe' in its wake through which the water drains away. A

Royal Commission in 1880 reported evidence that between 1846 and 1873 more than 10 million acres of farming land were drained by one means or another, and as one would expect, the greatest periods of activity coincided with the times when farming was generally at its most prosperous.

Modern field drainage practice is not substantially different from that of the last century, although we now understand a great deal more about the processes involved. Any underdrainage system is bound to be expensive, so care must be taken to design it properly. Typical drainage networks are illustrated in fig.

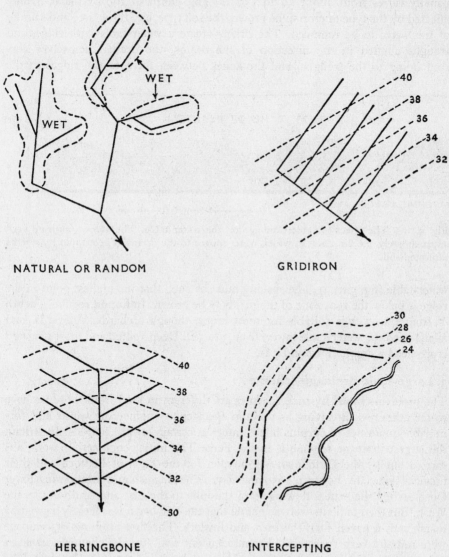

NATURAL OR RANDOM

GRIDIRON

HERRINGBONE

INTERCEPTING

Fig. 5.III.2 Some common types of drainage systems. (Adapted from Linsley and Franzini, 1964).

5.III.2. Standard underdrain pipes used to be about 2 in. in diameter, but it is now normal to use 4-in. ones. The distance between drains is controlled pre-eminently by the permeability of the soil to be drained. Where this is high, then the drains can be widely spaced, where low, then the drains have to be more closely spaced. In really heavy and sticky clay soils the density of drains might well have to be so high as to be too costly, in which case a mole plough might well be the answer. Gradients should not be less than 0·2%, to give a velocity of 1 ft/sec when flowing full, to avoid clogging by sediment. Spacing of drains usually varies from about 50 to 150 ft. The depth of the drains is mainly affected by the type of crop to be grown, the soil type, and the source and salinity of the water to be removed. The drains create a water-table with ridges and troughs aligned in the direction of the drains, the drains themselves cor-responding to the troughs, and the zones between them to the ridges in the

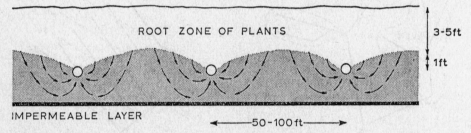

ROOT ZONE OF PLANTS   3-5ft

1ft

IMPERMEABLE LAYER   ←——50-100ft——→

Fig. 5.III.3 The effect of underdrainage on the water table. The arrows indicate very approximately the direction in which water moves to the drains. The stipple represents saturated soil.

water-table (fig. 5.III.3). The spacing must be such that the highest point in the ridge is below the root zone of the plants to be grown. In humid regions a depth of from 2 to 3 ft is suitable for most crops, though orchards, vineyards, and alfalfa need the water-table lower, from 3 to 5 ft. Deep-rooted and most irrigated crops need a depth of around 5 ft.

3. Large-scale drainage schemes

The problems posed by such schemes are different in kind from those we have so far examined, involving as they do spectacular engineering works and for-midable quantities of surplus flood waters at certain critical periods. In Britain the largest drainage scheme is in the Fens. The initial engineering work was carried out by the Dutchman Vermuyden for the Earls of Bedford and their financial associates between 1630 and 1652. The sluggish and meandering River Ouse carried the waters flowing from the hills to the east and south out to the Wash. But the gradients were so gentle that the Fen basin was largely a swampy marshland, a haven for fishermen and fowlers. The first attempts at drainage were initially very successful. Vermuyden cut two straight channels, each 21 miles long, between Earith Sluice in Huntingdon and Denver in Norfolk. The purpose of the sluice at Denver was to prevent the tides flowing back up the course of the Ten Mile River towards Ely, and that at Earith was to allow the engineers to decide in an emergency how much of the waters passing the sluice should be allowed to flow down the straight channels (the Bedford Levels) and how much down the original course of the Old West River (fig. 5.III.4). The two new channels reduced by 10 miles the length of the original course. By mani-

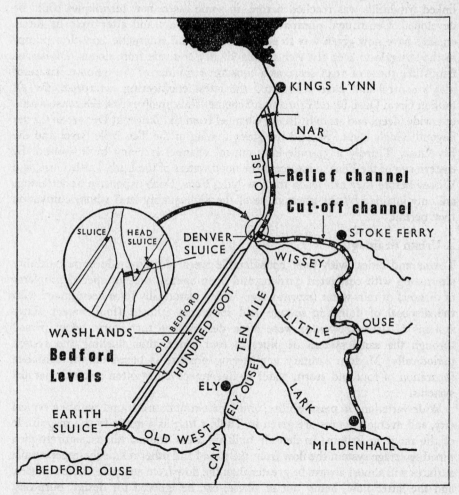

Fig. 5.III.4 The Fens. Drainage works.

pulating the two sluices, and by using the considerable storage capacity of the 'Washlands' between the two channels, the flood dangers could be minimized. The scheme was shortly to reveal unexpected difficulties. The peaty soils in the drained area very soon began to shrink as they dried out, and in addition bacterial action on the oxidized outer layers exposed to the air caused extra wastage. The loss of soil was less worrying than the fact that the level of the drained fields was steadily falling in relation to the rivers into which the waters drawn from the fields were drained. Soon it became necessary to use windmills to operate pumps, and before long, since the lifting power of pumps is limited, 'flights' of windmills linked in series became a common sight in the Fens. The shrinkage has continued. In Holme Fen (Hunts) there is a well-known cast-iron pillar which reveals a shrinkage of nearly 13 ft since 1848. Unfortunately the limit of series-linked windmills was reached before, in some cases, new techniques could be developed. Centrifugal pumps driven first by steam and after 1913 by diesel engines have now given way to electric motors and automatic axial-flow pumps in the struggle to keep the rich farmlands dry and safe from floods. Disastrous floods like those of 1947 serve as a constant reminder of the tenuous nature of man's control and have stimulated the latest engineering enterprise, the £8 million Great Ouse Flood Protection Scheme. This involves first the construction of a wide, deep, and straight Relief Channel from the sluices at Denver to the sea beyond King's Lynn. Second the straightening of the Ten Mile River and the Ely Ouse. Thirdly a 30-mile-long cut-off channel is being built around the eastern rim of the Fens to intercept the flood waters of the Lark, Little Ouse, and Wissey before they can reach the low-lying Fens. Such expensive undertakings are only justified by the great value of the high-quality land whose cultivation they permit.

4. Urban drainage

Towns and cities, with their considerable areas of man-made impermeability alternating with cultivated gardens and open spaces, present special problems of disposal of rain-water (storm water) which historically have been linked with the disposal of domestic sewage and industrial effluent (foul water). Early systems of urban sewerage were often designed so that storm water passed through the same system of pipes as foul water, thus flushing the system periodically. Modern sanitary engineering practice is based on the complete separation of foul and storm water in unconnected but often spatially parallel systems.

Wide variations in permeability, and hence runoff, are found within a typical city, and average figures are given in Table 5.III.3 as a guide to the magnitude of the range in Britain. In densely built-up areas drained almost entirely by a piped sewerage system the flow from the paved and other relatively impermeable surfaces will almost always be greater than the flow from any permeable surfaces, and the latter flow, being out of phase, can be ignored for design purposes. Where a significant part of the catchment area is not yet built upon and is still primarily drained by streams and rivers, the total runoff has a large component from the latter source. This secondary flow is generally assumed to reach its peak at a time equal to twice the estimated time of concentration used for the primary flow.

In designing an urban sewerage system the engineer has to consider not only the total quantity of water to be handled but also the time period over which a particular pattern of rainfall will be associated with the system. A storm, for instance, will fall with varying intensity over a catchment area, but different

TABLE 5.III.3 Impermeability factors, storm-water sewers, urban and sub-urban areas

Density (houses/acre)	Percentage impermeability	
	Primary flow	Secondary flow*
Industrial	50–100	12·5–0
Special areas	40–60	15·0–10
30	68	8
25	64	9
20	55	11
12	38	15·5
9	30	17·5
6	25	18·8
2 Rural	5–10	23·8–22·5
Open spaces	5	23·8

* Runoff coefficients vary according to sub-soil characteristics and cultivation. Typical values are:

    60–75% for rocky or clay sub-soil on steep slopes
    50–60% undrained clay sub-soil on steep slopes
    20–40% drained and cultivated clay sub-soil
    10–20% cultivated sandy or loamy sub-soil
    5–10% woodland area.

parts of the catchment will discharge into the sewers at different times, and water in the system will take time to pass through it. The storm water will obviously not all have to be handled at one point in time. The design of the system must be such that it will handle all that is presented to it under most, but rarely under all, circumstances. To make a system large enough to cater for even the exceptional 'once in a century' storm would be unnecessarily expensive, so a decision has to be made concerning the severity of the storms the system is going to be designed to handle. In Britain it is the sudden short-lived summer storm that throws the greatest strain on the storm-water sewers. The Road Research Laboratory hydrograph method is now generally regarded as the most suitable for calculating runoff from all but the smallest urban areas (i.e. less than 20 acres of impermeable surfaces). In essence five steps are involved, illustrated in fig. 5.III.5. First, the catchment area is divided into sub-areas of a convenient size. Second, the impermeable area of each sub-area is calculated and an area–time diagram drawn for each. The area–time diagram relates the area contributing to the rate of flow with the time after the start of the rainfall. The graphs for each contributing area are combined to form a cumulative total profile for the whole system (fig. 5.III.5(b)). Third, a rainfall profile is drawn from tables prepared by the Road Research Laboratory. These are based on long-term observations and show the minute-by-minute changes in the amount of rain received during storms of varying severity and frequency. Profiles for storm frequencies

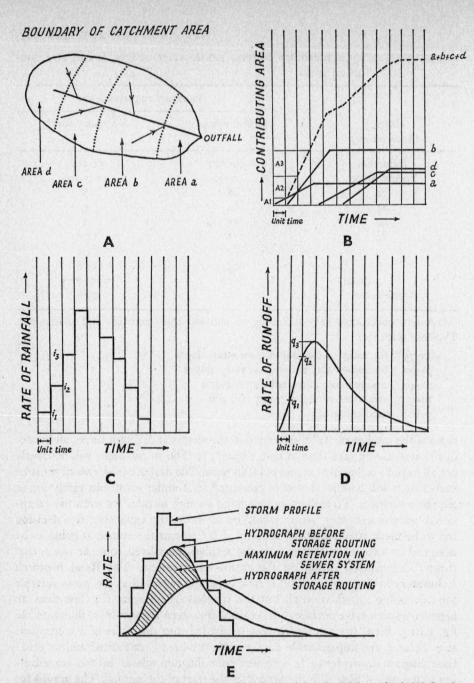

Fig. 5.III.5 Procedure for calculating runoff hydrograph by the RRL method. (A) Catchment area with four sub-areas; (B) Area/time diagram; (C) Rainfall profile; (D) Runoff hydrograph before allowing for storage, $q_1 = (i_1 \times A_1)$, $q_2 = (i_1 \times A_2) + (i_2 \times A_1)$, $q_3 = (i_1 \times A_3) + (i_2 \times A_2) + (i_3 \times A_1)$; (E) The effect on the hydrograph of allowing for retention in the sewer system (After Watkins, 1966. By permission of The Controller of H.M. Stationery Office. Crown copyright reserved).

expected once a year, once in five, once in ten, and once in thirty years are available. A simplified and generalized profile is shown in fig. 5.III.5(c). Fourthly, the two profiles already determined, the cumulative total area–time profile and the appropriate rainfall profile, are combined to give the first hydrograph (fig. 5.III.5(d)). This, however, takes no account of the storage capacity of the system itself. Methods are available for calculating the effects of the additional factor and for determining the form of the final hydrograph. This fifth step is illustrated in fig. 5.III.5(e). In practice, the method gives excellent results, and the number of calculations required make it particularly suitable for solution by digital computer.

Most work on urban sewerage systems involves redesigning existing systems to accommodate changes in input or to remove operational difficulties. A typical problem is illustrated in fig. 5.III.6. The area in question is one of fairly substantial Victorian villas with some more recent development. There are a number of open spaces, and gardens are generally large for an older urban area. The storm- and foul-water systems are inadequate, and both need redesigning. The first map shows the impermeable areas, i.e. those of slate, stone, brick, concrete, stucco, asphalt, etc. The next map shows the way in which the entire area has been subdivided into smaller units, the delimitation being based on the natural direction of movement of surface water (the gentle regional slope is to the outfall in the north-west of the area), and also the alignment of the existing sewerage system. The dotted areas on this map show the areas at present liable to flood. The third map shows the sketch of the sewerage system, defining the main lineaments of the network. Each link is numbered (decimal point o always indicates the length of sewer most upstream of the outfall), and the thicker line and the prefix 1 represents the trunk sewer, having its outfall at the end of link number 1 : 14. Once data about each catchment area have been compiled along the lines indicated, a standard computer programme is available very rapidly and therefore cheaply to calculate the expected flows along each link and the size of pipe needed to accommodate them. In the example quoted the final form of the system is shown in fig. 5.III.6(d).

In most large cities alterations to the sewerage systems involve adding extra links to the existing network or selectively increasing the capacity of the system, so the freedom for fundamental redesign is severely limited. In a new town, however, the possibility, at least in theory, exists for using computer methods to optimize not only the operational characteristics of the sewerage system but also to overall design.

REFERENCES

DEPARTMENT OF SCIENTIFIC AND INDUSTRIAL RESEARCH [1963], *A Guide for Engineers to the Design of Storm Sewer Systems;* Road Note No. 35 (H.M.S.O., London).

FUSSELL, G. E. [1952], *The Farmers' Tools;* (Andrew Melrose, London).

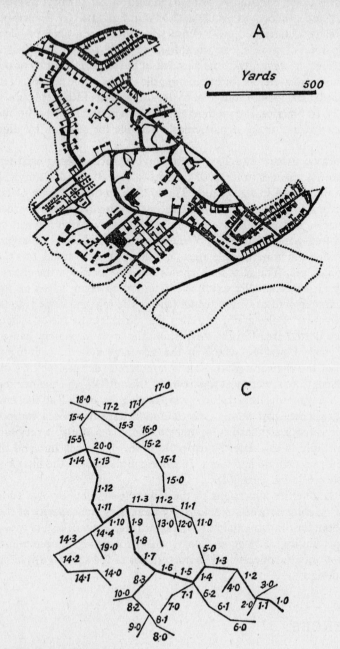

Fig. 5.III.6 Storm-water sewers.

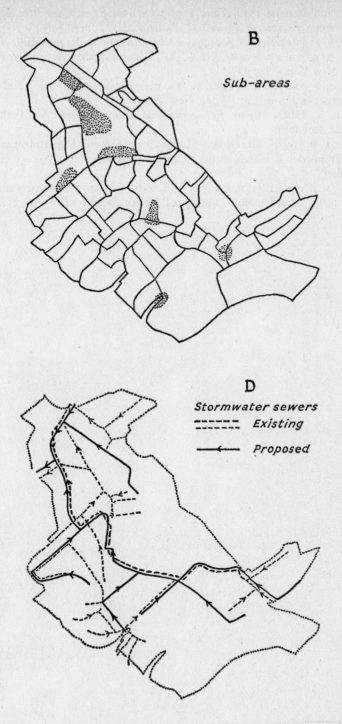

B

Sub-areas

D

Stormwater sewers

- - - - - - *Existing*

———←——— *Proposed*

HOUSE OF COMMONS, 1836, *Third Report from the Select Committee Appointed to Inquire into the State of Agriculture;* Parliamentary Papers, July.

KOHNKE, H. and BERTRAND, A. [1959], *Soil Conservation* (McGraw-Hill, New York).

LINSLEY, R. K. and FRANZINI, J. B. [1964], *Water-Resources Engineering* (McGraw-Hill, New York), 654 p.

SCHWAB, G. O., FREVERT, R., EDMINSTER, T., and BARNES, K. [1966], *Soil and Water Conservation Engineering* (Wiley, New York).

THORN, R. B., Editor [1966], *Engineering and Water Conservation Works* (Butterworths, London).

WATKINS, L. H. [1962], *The Design of Urban Sewer Systems;* Department of Scientific and Industrial Research, Road Research Technical Paper No. 55 (H.M.S.O., London).

WOLMAN, M. G. and SCHICK, A. P. [1967], Effects of construction on fluvial sediment, Urban and suburban areas of Maryland; *Water Resources Research*, 3 (2), 451–64.

6.III. Human Use of Ground Water [1]

R. L. NACE
U.S. Geological Survey

1. The fountains of the deep

The plains of the Iranian Plateau are dotted at places with thousands of circular earthen mounds surrounding vertical shafts as much as 200 m deep. These are access shafts to nearly horizontal tunnels, called khanats (fig. 6.III.1). The head segment of each tunnel intercepts and cuts below the water-table in permeable

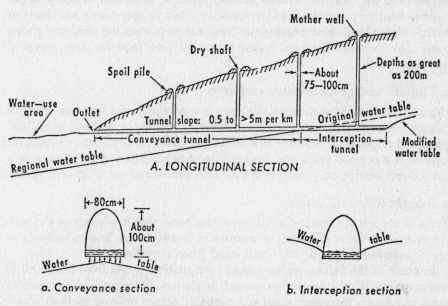

A. LONGITUDINAL SECTION

a. Conveyance section

b. Interception section

B. TRANSVERSE SECTIONS

Fig. 6.III.1 Cross-sections of a typical khanat (After Massoumi, 1966).

gravel. Downslope, the grade of the tunnel is less steep than that of either the land surface or the water-table. Where the tunnel reaches the surface, water can be led from it by canal to towns or irrigated fields.

About 25,000 khanats exist in Iran today, the longest being about 70 km. Their origin is lost in antiquity, but as early as 714 B.C., when King Sargon II of

[1] Publication authorized by the Director, U.S. Geological Survey.

Assyria invaded Armenia he found khanats there (De Camp [1963], p. 66). These he destroyed but he took the idea back to Assyria, whence it spread through the East, North Africa, and as far away as China. Much later, Spaniards introduced the khanat to Chile, and some are still in operation on the Atacama Desert, where they are known as *socavónes* (Dixey [1966], p. 91).

Many other examples are available of effective exploitation of ground water since antiquity in dry areas. The biblical characters who spoke of the fountains of the deep knew whereof they spoke. Wells, ground-water interception and diversion tunnels, enlargement of springs, artificial recharge, all date from antiquity. Only the technology for exploitation has improved.

The uses for ground water are far more varied than merely for domestic supply or irrigation. The spectacular geysers of Yellowstone National Park, which are natural hot-water fountains, have thrilled millions of tourists. Springs that discharge high on the face of a cliff in southern Idaho are led through a penstock to generate hydro-electric power. Icelanders use volcanically warmed water for space heating. In Italy steam from hot ground water drives electrical generators. Mineralized and thermal springs are the principal attraction of many health resorts. Underground pools and rivulets add zest to spelunking and harbour strange biota that fascinate biologists. Some fish hatcheries use clear cool spring water. For centuries the rural spring house has been used for cool storage of perishable food.

2. Climate and ground-water recharge

Climate has fluctuated through the ages, but no significant one-way trend or change has occurred during recent millennia. Historical, archaeological, and geological evidence all converge to indicate that climate today is substantially the same as it was 8,000 years ago. Recognition of this fact is important because it has a direct bearing on problems of the development and use of ground water.

3. Inherited ground water

Some of the world's greatest aquifers contain water that is a legacy of the past. An example is the vast system of aquifers in North Africa, known variously as the Nubian sandstone and the Continental Intercalary Fountain.

In much of the Sahara region annual precipitation ranges from nearly nil to about 250 mm. Few areas receive as much as 500 mm. Owing to high evaporation rates, recharge is extremely small and probably occurs only where flash runoff from intense local storms reaches outcrops of the aquifer. Estimated total storage in the aquifer beneath 6.5×10^6 km^2 of area is of the order of 600×10^3 km^3 (Jones, 1965). Current recharge is relatively negligible, and it is evident that most of the water entered this aquifer during and before the pluvial period at the end of the Pleistocene epoch. Radio-carbon analyses of water samples indicate that some of the water entered the aquifer 30,000–40,000 years ago.

Another example is the Ogallala formation, a sand and gravel aquifer in the High Plains of Texas and adjacent States. South of the Canadian River in Texas and east of the Pecos River in New Mexico, an area of about 90,000 km^2 is isolated from any source of replenishment by underflow. No perennial streams flow through the area and none rise within it. Streams that rise in the eastern area receive but little water by natural discharge from the aquifer. Because of low rainfall, high evaporation, and low permeability of subsoil, estimated recharge from direct infiltration of precipitation is only about 10–15 mm. Some estimates are as low as 4 mm.

The estimated water content of the aquifer in 1938, before the onset of heavy pumping, was about 600 km³, of which about half is theoretically recoverable. Calculations based on underflow velocities and distances through the aquifer indicate that some of the water has been in transit during at least 13,000 years.

Pumpage from the Ogallala formation has reduced storage by nearly 110 km³, and annual withdrawals in recent years have been about 6·2 km³. This is about six to fifteen times the variously estimated natural recharge rates, so the reserve is being depleted and could be exhausted in the foreseeable future. Complete exhaustion, of course, will not actually occur. Declining water levels will lead to lower yields, higher pumping lifts and higher costs, and operations that become uneconomical will end; thus, economics will force a balance between supply and draft. This has already happened in some parts of the High Plains.

Many parts of the world contain inherited ground water in smaller but important aquifers. The water in such aquifers may be considered as non-cyclic because it would not naturally participate in the water cycle within a humanly significant span of time. For example, a sedimentary aquifer in the vicinity of Maracaibo, Venezuela, is about 1,300 km² in extent and has a water-storage volume of 35 km³. Radio-carbon analyses of samples of the waters indicate that they range in age from 4,000 to 35,000 years.

4. Cyclic ground water

The well-known characterization of aquifers as both reservoirs and pipelines is appropriate, but the pipeline analogy is less apt than that of the reservoir. In fissured limestone with openings so large that Darcy-law flow does not occur the speed of ground-water movement may be comparable to that of some rivers at low stage – about 0·03–0·3 m s⁻¹. In cavernous limestone speeds may be as high as 7–10 m s⁻¹ (Pardé [1965], p. 38). The pipeline analogy is apt for such aquifers.

On the other hand, a study of ground-water movement in finely fractured crystalline rock (Marine, 1967) indicated an average speed of about 17·5 × 10⁻⁶ m s⁻¹ (about 1·5 m da⁻¹). Even slower speeds prevail in some argillaceous sediments. Here the pipeline analogy becomes strained. Well-sorted clean gravel is excellent material, but 'all the water moving through clean, well-sorted gravel in a bed 100 ft thick and 1 mile wide [30 m and 1·6 km] with 1% gradient could be transmitted in a pipe 14 inches [about 35 cm] in diameter at the same gradient' (Thomas and Peterson [1967], p. 71). On the other hand, water stored in 1 square mile (about 4 km²) of the same aquifer could supply the same discharge pipe for about 4 years.

These few data illustrate that though ground water is generally in continuous motion, its detention time in aquifers is long compared to detention of water in river channels. Hence, great apparent age of a ground-water sample is not in itself an indication that recharge is small or that the water is non-cyclic. A study in the Great Artesian Basin of Queensland and New South Wales, Australia, indicated that water in the aquifers 30–80 km from the recharge areas has been underground about 20,000 years (Water Research Foundation, 1964). This implies a travel speed of 1·5–4 m yr^{-1}.

In many aquifers, however, underflow speeds range from a fraction of a metre to several metres per day. Water-detention times range from a few minutes to a few decades. In extensive highly permeable aquifers such as the basalt of the Snake River Plain in Idaho, occupying 30,000 km^2, underflow speeds average 10–15 m da^{-1}, and detention periods range up to several hundred years.

5. Management of ground water

Management of water means controlled use in accord with some plan. Most uses of ground water are simple exploitation as a free good, and the water is generally considered to be self-renewing. Preceding remarks show, however, that recharge of some aquifers is so small or slow that renewal will not occur naturally within spans of time that are relevant to current planning or management. Moreover, many aquifers are so inefficient as pipelines that relatively little of their recoverable water can be extracted where and when it is wanted.

Discharging wells do not lower the water-table or pressure surface uniformly throughout an aquifer. In an aquifer like the Ogallala formation, for example, the water-table is not flat but dimpled, each dimple being a cone of depression around a well. The intervening mounds are in segments of the aquifer that have not been dewatered. Springs and seeps along the natural discharge area of the Texas aquifer will continue to discharge. All pumpage within the plain, therefore, is a depletion draft on reserve storage because pumping of wells will not reduce natural discharge until pumping effects reach the discharge area and lower the hydraulic gradient there. The only measure which will permit pumping from the Ogallala at present rates to continue into the far future is artificial recharge with imported water. The area lacks good surface reservoir sites, but the ground-water reservoir has tremendous unused storage capacity.

Israel is an example of a modern industrial–agricultural economy where water is so scarce that total management of ground water and surface water as a unit resource is practised. This is necessary in order to get the most out of every drop of water and to control sea-water incursion in coastal aquifers. Even humid areas like Western Germany and the Netherlands have had to establish careful control, because increasing pollution of all waters makes it increasingly difficult to produce suitable water for growing demands.

Where water demand is large, surface-water supplies, if available, have generally been exploited first, because the water is visible, the amount available can be measured readily, and run-of-the-river diversions are cheap and require no great engineering works. Ground water generally has been quite independently exploited, as though it were a different or separate resource. In fact, however, ground water and surface water typically are in hydraulic continuity,

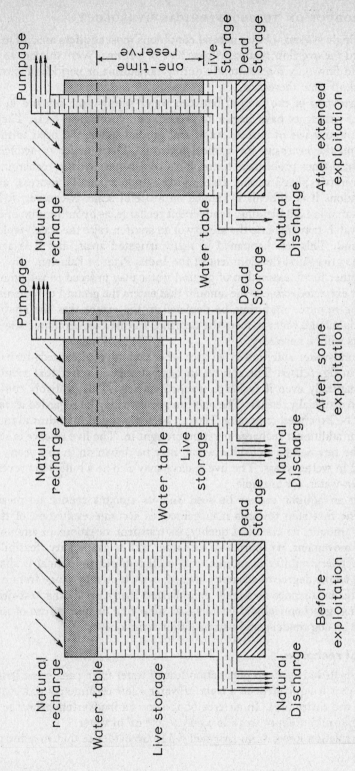

Fig. 6.III.2 Stylized representation of planned exploitation of ground-water storage in order to capture a larger share of current water yield (After Mandel, 1967).

parts of a single system. Under natural conditions most aquifers are full to overflowing, and the overflow maintains base flow of streams. Were this not so, most rivers would flow only intermittently during rainstorms or periods of snowmelt and for a short time thereafter.

With narrowing of the margin between water supply and demand in many areas, water managers have begun to accept the hydrological facts. They talk about conjunctive use of surface water and ground water, the need to manage aquifers, and the necessity for artifical recharge. The practice of artificial recharging dates from prehistoric times, but extensive systematic recharging is a twentieth-century phenomenon. Recharging is not a simple process, and in many situations it is difficult to induce on a useful scale. Moreover, it is not necessarily an unmixed blessing. Inadvertent recharge, as by infiltration of excess irrigation water, may increase the salinity of an aquifer, raise the water-table, and waterlog land. This has happened in many irrigated areas, and it is a major problem in agricultural development of the Indus Plain in Pakistan.

On the other hand, extraction of ground water may proceed to the point that the amount extracted exceeds the amount that enters the ground to replenish the supply. The resource may thus be depleted to the extent that the aquifer no longer yields enough water to meet the demand. In that case economic and social adjustments will be necessary.

The ground-water situation in an area where withdrawal exceeds recharge is illustrated in fig. 6.III.2. The reserve in dead storage is water that would not discharge naturally even if recharge stopped entirely. This probably could not be extracted artificially, either. The one-time reserve may be regarded as capital, which may be expended only once. Having been expended, withdrawal must be decreased or additional recharge must be brought in. The live reserve is storage that must be retained as a cushion which may be drawn on in dry years, to be replenished in wetter years. The live reserve may also be a buffer to prevent invasion of sea-water, for example.

Water in an aquifer cannot be seen and its amount cannot be measured directly. The first step towards management is accurate evaluation of the resource: Its amount, its chemical quality, its transient variations in amount and quality, its movement, its sources of recharge, and its availability (feasibility of extraction). Every aquifer is unique, and each requires individual evaluation. Depending on the degree of accuracy needed, evaluation may range from hydrogeological reconnaissance to highly complicated study, including test-drilling, test-pumping, geochemical and geophysical surveying, mapping, use of mathematical and analog models, and various other techniques.

6. Artificial recharge

In many irrigated areas deep percolation loss of water from canals and irrigated fields each year is equivalent to a layer of water a few centimetres to a metre in depth over the entire area. In an area of 100,000 ha inadvertent recharge from irrigation thus may amount to 30 to $1,000 \times 10^6$ m^3 of water.

Deep percolation losses from irrigated fields are evidence that in some places recharge is easy to induce. In situations where recharge amounts to a loss of useful water or an aggravation of water-logging, the rate seems high. Where recharge is desired, however, infiltration rates of a few centimetres or metres per year are low. Generally, it is not feasible to inundate tens of thousands of hectares for the sole purpose of recharge. Evaporation loss would be high, and extensive flat areas generally are too valuable for other purposes. Recharging is usually practicable only where the unit-area rate of recharge is high or where water can be injected underground through intake wells. The sands and gravels of Long Island, New York, and of the alluvial fans around the margins of the Central Valley of California are examples of favourable areas for water spreading.

The High Plains of Texas and New Mexico exemplify conditions less favourable for recharge. The plains are dotted by many hundreds of playa basins that fill with water from local runoff during intense storms. The subsoil is permeable, but the basins are largely floored by fine silt, and much of the area is underlain by poorly permeable caliche. Therefore, depending on local conditions, 20–80% of the ponded water evaporates rather than percolating underground (Havens [1966], p. 35). In order to reduce this loss, experiments have been widespread to get the water underground through intake wells. The degree of success has ranged from poor to excellent during short periods. Suspended sediment in the injected water clogs the aquifer, and the wells must be pumped periodically to remove the sediment (Myers, 1964). After a few weeks or months the wells may cease to be effective.

In the Grand Prairie rice-growing region of Arkansas the water supply would be ample if ways could be found to store excess winter streamflow. This could be accomplished by injecting the water underground, but clogging by sediment is a problem, and chemical reactions between the natural ground water and the injected water produce precipitates. Entrained air and micro-organisms also cause clogging (Sniegocki, 1963). In order for recharge to be successful, injected water must be pretreated essentially as though for a municipal water supply. The problem thus is one of economics.

Even at places where recharge is feasible, the inefficiency of aquifers as pipelines again is a drawback. Physical transit of water is not the whole story, because transfer of hydraulic head is more rapid. Accessions by recharge create a so-called recharge wave, and this wave may travel at rates of a fraction of a metre to several hundred metres per day, depending on permeability of the aquifer. So far as water availability is concerned, the effect is the same as physical transfer of water. However, in some aquifers transfer of head is so slow that injected water, for practical purposes, is recoverable only in the vicinity of the injection site.

7. Pollution

All natural water is contaminated in the chemical sense, in that it contains substances other than H_2O. Pollution, on the other hand, is a totally subjective concept, and few definitions are satisfactory. In general, water managers consider that water is polluted when dissolved or entrained substances are present in amounts that make the water unfit or undesirable for specified uses. To the hydrologist, however, *polluted* is an incongruous label for a natural water, which he would call mineralized.

Absence of serious pollution of ground water in some places where sources of pollution are widespread is a consequence of the nature of aquifers. A sand aquifer acts like a sand filter in a water-treatment plant. By sorption processes it can also function as an ion-exchange bed. Conditions within aquifers are largely anaerobic, and lack of oxygen kills many waterborne organisms.

On the other hand, ground water will not necessarily purify itself through any given distance or period of time. This is especially true of aquifers such as cavernous limestone, which carry organic and mineral pollutants rapidly through considerable distances. The generally slow motion of ground water in granular aquifers is highly important in relation to pollution. Once polluted, an aquifer may remain so for years or centuries, because flushing action is very slow. Studies of industrial waste in the ground water of Long Island (Perlmutter, Lieber and Frauenthal, 1963) show that, even if pollution were discontinued, at least ten years would pass before all the polluted water would be discharged from the aquifer.

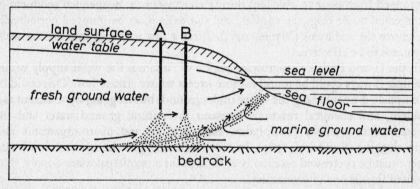

Fig. 6.III.3 Cross-section showing dynamic relation between salt water and fresh in a coastal aquifer. The shaded area represents the zone of transition. The deepest shading represents the zone where the horizontal gradient is nil. (Adapted from Kohout and Kiein, 1967).

Salt-water invasion of aquifers is a growing problem in most areas near the sea or other bodies of salt water. The Ghyben-Herzberg concept of a lens of static fresh water floating on salt water beneath a circular oceanic island is inadequate to describe the situation and processes in coastal areas.

Figure 6.III.3 is a simplified portrayal of a coastal situation. The interface between salty and fresh water is actually a zone of transition from fresh to salt water. Motions of these waters are indicated by curvilinear flow lines. In the transition zone seaward-flowing fresh ground water dilutes and sweeps out the intruding sea-water. The width of the zone is controlled by permeability of the aquifer and velocity of flow, and by the range of oceanic tides. Heavy pumping of the well at A may create an upward tongue of salt water which may reach that well itself and may also migrate to other wells, as at B.

The actual situation in water-table aquifers and in limestone and artesian aquifers along coasts is much more complicated than that in the simplified portrayal. Each aquifer and locality is, in fact, a special case, and each must be studied individually (see Cooper [1959] and Kohout and Klein [1967]).

Acknowledgements. The author gratefully acknowledges constructive review of this chapter and suggestions by his colleagues, J. H. Feth, Howard Klein, B. N. Myers, and A. G. Winslow.

REFERENCES

COOPER, H. H. [1959], A hypothesis concerning the dynamic balance of fresh water and salt water in a coastal aquifer; *Journal of Geophysical Research*, **64**, 461–7.

DE CAMP, L. S. [1963], *The Ancient Engineers* (Garden City, New York), 409 p.

DIXEY, F. [1966], Water supply, use and management; *In* Hills, E. S., Editor, *Arid Lands, a geographical appraisal* (London and Paris), pp. 77–102.

HAVENS, J. S. [1966], Recharge studies on the High Plains in Northern Lea County, New Mexico; *U.S. Geological Survey Water Supply Paper* 1819-F, 52 p.

JONES, J. R. [1965], Written communication to R. L. Nace, 22 October.

KOHOUT, F. A. and KLEIN, HOWARD, [1967], Effect of pulse recharge on the zone of diffusion in the Biscayne aquifer; *International Association of Scientific Hydrology Publication*, **72**, pp. 252–270.

MANDEL, S. [1967], Underground water; *International Science and Technology*, **66**, 35–41.

MARINE, I. W. [1967], The use of a tracer test to verify an estimate of the groundwater velocity in fractured crystalline rock at the Savannah River Plant near Aiken, South Carolina; *American Geophysical Union, Geophysical Monograph Series*, **11**, 171–9.

MASSOUMI, AHMAD, 1966, *Groundwater production in Iran:* Paper presented to United Nations Seminar on Methods and Techniques of Ground-Water Investigation and Development, Tehran, October 16 (duplicated), 12 p, 5 figs.

MYERS, B. N. [1964], Artificial-recharge studies; In Cronin, J. G., editor, A summary of the occurrence and development of ground water in the Southern High Plains of Texas; *U.S. Geological Survey Water Supply Paper* 1693, 56–71.

PERLMUTTER, N. M., LIEBER, M. and FRAUENTHAL, H. L. [1963], Movement of waterborne cadmium and hexavalent chromium wastes in South Farmingdale, Nassau County, Long Island, New York; *U.S. Geological Survey Professional Paper* 475-C, 179–84.

PARDÉ, M. [1965], Influences de la permeabilité sur le régime des riviéres; *Colloquim Geographicum*, **7**, 21–100.

SNIEGOCKI, R. T. [1963], Problems in artificial recharge through wells in the Grand Prairie region, Arkansas; *U.S. Geological Survey Water Supply Paper* 1615-F, 25 p.

THOMAS, H. E., and PETERSON, D. F., JR [1967], Ground water supply and development; In Hagan, R. M. and others, Editors, Irrigation of agricultural lands; *American Society for Agronomy, Agronomy Series*, **11**, 70–91.

WATER RESEARCH FOUNDATION OF AUSTRALIA [1964], Nuclear chemical study of the age and renewal rate of the Great Artesian Basin; *Water Research Foundation 9th Annual Report and Balance Sheet*, 12–13.

7.III(i). The Human Use of Open Channels

ROBERT P. BECKINSALE

School of Geography, Oxford University

The human use of rivers and streams depends on the nature of the rivers as well as on the needs and customs of the riverine societies. Most modern methods of river use are merely the technical refinements of usages practised in some advanced early civilizations long before the Christian era.

Today the problems of the human use of open channels with regimes fall under nine main interrelated facets: flood control; irrigation and drainage; water power; flotability; navigation; water supply; fishing and wild-life conservation; recreation and religion; and water-pollution (fig. 7.III(i).1). In the following short discussion, together with Chapter 10.III, which treats the first three facets most dependent upon the regime variations of discharge, little more can be attempted than to demonstrate these aspects from notable examples so as to illustrate the present state of man's response to the socio-economic opportunities offered by rivers.

1. Flotability

The crudest human use of rivers, the uncontrolled floating of objects, is still very popular. In large areas the timber, pulp, and paper industries depend mainly upon river floatways for the transport of their raw material. This is especially so in the northern forests of Canada, Scandinavia, and the U.S.S.R., where the river regimes (D) have a spate in spring or early summer. Many early sawmills grew up inland near waterfalls, and in, for example, Norway a few mills still use direct water-power. With the coming of steam-power and later of hydroelectricity the chief pulp-processing mills tended to be sited at the junction of rivers or at their mouths in more convenient locations for markets and export. Because typical extra-tropical timbers float easily, it is usual in many countries to pile the logs in winter near floatways, into which they can be rapidly pushed at the thaw. As in much of Scandinavia the thaw progresses inland and upward, a fairly continuous supply of logs can be sent to the river mouth in the floating season. In most lumbering regions various forms of mechanical transport are used to take logs to the mill, but the prime method is still to drag logs to a floatway. In Norway in 1960 nearly 10,000 miles of floatway handled 33 million logs; in Sweden a public-floatway system of over 21,000 miles, with a staff of nearly 40,000 persons, deals with about 170 million logs annually, and the number handled has risen to 208 million (in 1949). In Finland about 25,000 miles

Fig. 7.III(i).1 Model for multiple-purpose integrated river basin development.

1. Multiple-purpose reservoir.
2. Recreation; swimming, fishing, camping.
3. Hydroelectric station.
4. Municipal water supply.
5. City and industrial waste treatment plant.
6. Pump to equalizing reservoir for irrigation.
7. Diversion dam and lake.
8. High-level irrigation canal.
9. Levees for flood control.
10. Erosion control: stream dams and contour terracing.
11. Regulating basin for irrigation.
12. Wildlife refuge.
13. Low-level irrigation canal.
14. Gravity irrigation.
15. Contour ploughing.
16. Sprinkler irrigation.
17. Community water treatment plant.
18. Navigation: barge trains, locks, etc.
19. Re-regulating reservoir with locks.
20. Farm pond with pisciculture.

(Adapted from *A Water Policy for the American People*, Vol. I, X)

of floatway are in use, and the busiest carry up to 3 million m³ of timber annually for the national production of nearly 4½ million tons of pulp and 1½ million tons of paper (fig. 7.III(i).2).

The mass floating of logs causes problems for hydroelectric operators;

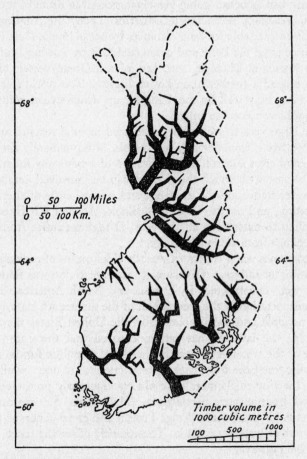

Fig. 7.III(i).2 Chief timber floatways in Finland, 1960.
The blank areas in parts of the south-west occur largely because here short-haul road transport has replaced river floating (Adapted from *Atlas of Finland*, 1960 and Millward, 1965).

similarly, dams hinder log floating and complicated legal disputes have been common. In Finland the eventual solution was to compel the hydrosite constructor to build a by-pass trough for logs and to compensate the log-floating company for 1½ times its direct material losses caused by damming the river. Difficulties of another kind occur in the tropics, where many valuable hardwoods do not float when green. Teak is felled some time before being dragged to the waterfront, a task often performed in Burma by elephants.

2. Navigation

Treaties concerning navigation on the River Po date back to A.D. 1177 and on the Rhine to A.D. 1255. For thousands of years there was little difference between sea and river craft, but as ocean-going vessels increased in draught their navigation inland was increasingly restricted to estuaries. Today the Amazon is the only unaltered waterway usable for long distances by ocean liners. Here floating docks a short distance from the bank and connected to it by a swing bridge suffice for ocean-going vessels at Manaus, 2000 km inland. Ocean vessels of up to 14 ft draught can ascend a further 1700 km to Iquitos. The world's second longest natural waterway may well be the Yangtze, up which ocean-going vessels can ascend to Hankow, 1,000 km inland.

Elsewhere large vessels are usually restricted to tidal reaches and lakes, and navigation upstream, even for smaller vessels, often demands dredging, locks, canalization, and river control. The utilization of a waterway for traffic depends heavily on the ease with which the channel can be improved, on its connections with other water routes, on the existence and relative costs of alternative means of transportation, and on the freights available. AF and CF river regimes are especially suited to navigation all the year. D regimes suffer from seasonal ice and other regimes from seasonal drought.

In many countries water transport was the chief means of moving heavy goods until the coming of railways. Occasionally steamer traffic was flourishing when the railways were constructed. In Eurasia and North America the survival of waterway transportation in the second half of the nineteenth century was often a question of national and social policies. In the United States the Constitution, and since 1824 the laws on navigation, ensured that the improvements and maintenance of the waterways were financed out of public funds. All navigable waterways were free from tolls, with no restrictions as to user. Similarly, in more recent times the Soviet planners have always vigorously promoted river transportation and hydroelectric installation. They have, for example, allocated enormous funds for the Greater Volga Project and given generous tariff inducements in favour of waterway traffic. These tariffs offset the more direct routes followed by the railways.

Today river transport is especially important in industrialized countries with bulk cargoes and in some densely peopled lands, such as China and South-East Asia, where a large population of boat owners, often living partly or entirely on the water, combine habitations with trading or fishing. The latter seems to lack statistics; the former may be best exemplified from North America and Europe.

In the United States the minimum depth of 6 ft was increased to 9 ft in 1930, and since 1944 was raised to 12 ft on the lower 1,200 km of the Mississippi. By 1947 about 28,000 miles of the 65,000 miles of so-called navigable waterway in the country had been improved, and the waterway movement totalled 31,500 million ton-miles on the rivers and 112,000 million ton-miles on the Great Lakes. In 1964 the corresponding figures were 144,253 million ton-miles on rivers and 105,912 million ton-miles on the Great Lakes. The most impressive increase was on the Mississippi system, including great tributaries such as the Ohio. Here, in spite of competition from pipelines, the freight carried expanded from 115 million short tons (30,382 million ton-miles) in 1950 to over 226 million short tons (80,087 million ton-miles) in 1964. Much of this increase was due to improvements in navigability and in barge techniques.

Navigation on the Great Lakes is closed by ice for about four months annually, but it has long been the world's busiest system of inland water transport and is unsurpassed for size of boats and length of hauls. After 1932, when the Welland Canal (by-passing Niagara Falls) was reconstructed, lake carriers of 20,000–25,000 tons drawing 27 ft could navigate between all the Great Lakes. The Saulte-Ste-Marie canal between Lakes Huron and Superior was regularly used by over 110 million tons of cargo annually. But rapids on the lower St Lawrence closed the system for ocean-going traffic. Between 1953 and 1958 this was corrected by the construction of the St Lawrence Seaway, whereby a series of locks, connected with hydroelectric projects, lifts vessels 69 m to Lake Ontario. Thence the existing Welland Canal, which is considered a part of the Seaway, lifts vessels a further 100 m to Lake Erie. Lake carriers of up to 25,000 tons now carry iron ore from the lower St Lawrence to upper lake ports, and ocean vessels of up to 8,000 tons travel 2,700 km inland. Chicago, at nearly 180 m above sea-level, and Duluth-Superior now each deal direct with over $4\frac{1}{2}$ million tons of foreign cargo annually, irrespective of their large lake trade. Traffic on the Seaway, as distinct from that on the four upper lakes, rose to 67 million short tons of cargo in 1966, and of this amount 14 million tons came from overseas. The chief commodities by volume were iron ore, wheat, bituminous coal, corn, manufactured iron and steel, fuel oil, barley and soya-beans, all ideally suited to bulk handling.

In Europe, which is about the same size as the mainland United States, navigation is favoured by the low-lying plain that extends from the Netherlands to the Urals, but is hindered by political fragmentation. The busiest waterways are the Rhine, Elbe, Danube, and Volga. The Rhine takes 4,000-ton barges to the Ruhr and 2,000-ton boats as far as Basel, while many of its tributaries and feeder canals take barges of 1,000–1,500 tons. In 1965 cargo handled on the Rhine amounted to 223 million tons or a nearly 13% above that in the former peak year 1961. The international freight consisted of 16,050,000 tons carried on the Netherland's section; 79,599,000 tons (27·5 millions downstream and 52 millions upstream) over the German–Dutch frontier; and 12,310,000 tons between West German and French ports and Basel (Switzerland). To these amounts must be added the internal traffic of 59,048,000 tons between West German ports, and 56 million tons between Netherland's ports. The total freight movement amounted to 43,821 million ton-kilometres. Table 7.III(i).1 shows the length of navigable inland waterway (river and interconnecting canal) and certain freight characteristics of various countries in Europe in 1966. In the last column, A relates to all freight journeys while B excludes freight carried by short road hauls.

Except in Britain, in 1965 and 1966 the growth of water traffic in Europe was well maintained and was not seriously affected by new pipeline developments for crude petroleum. At Strasbourg, for example, decrease in crude-oil traffic was offset by the rise in the transport of refined petroleum products. The upper Rhine trade to and from Basel reached an all-time record of 8·8 million tons in 1965, while the cargo traffic on the newly finished canalized Moselle amounted to 3·4 million tons. The Moselle is being improved upstream of Metz, and by the early 1970s will be connected with the Saône, so providing a Rhine–Rhône waterway.

TABLE 7.III(i).1

Country	Total tonnage (1,000 tons)	Total ton/km (millions)	Average length of haul (km)	Total length of navigable waterway in regular use (km)	Percentage total natural freight move-ment carried by waterway (ton/km) A	B
Belgium	79·6	5,392	68	1,595	26*	
France	93·5	12,652	135	7,677	8	9
West Germany	207·9	44,826	216	4,496	26†	30
Netherlands	197·8	25,240	128	6,044	64	
U.S.S.R.	279·0	137,582	493	142,700	6	
United Kingdom	7·8	195	25	1,345	0·2	

* Percentage as for 1961. † Percentage for 1965.

The Danube, which traverses Alpine Europe, is being extensively regularized. The projects include thirteen dams with hydroelectric installations on the Austrian section alone, and a large barrage at Sip below the Iron Gates which will dam back the water far upstream of the rapids, and will provide locks 300 m long and 34 m wide for navigation. Since 1932 a Rhine–Danube link has been undertaken by a company with a State concession to exploit the local water-power and use the net earnings to finance canal building. By 1967 about forty-three hydrostations with a total annual production of 2,300 million kWh had been constructed in the Main basin, and their earnings had facilitated loans sufficient to complete the first 20 miles of the 120-mile Rhine–Danube connection. Boats of 1,500 tons can now reach Bamberg on Main, and will, it is planned, reach Nuremberg by 1970 and Regensburg on the Danube by 1981. Then boats of that size will be able to travel 2,125 miles from Rotterdam to the Black Sea, and tonnage at Danube ports (21 million tons of cargo loaded and 22·5 million tons unloaded in 1962) will greatly increase. The freight carried on the river in 1966 weighed 45 million tons, of which 28 million was internal traffic.

The third great expansion of the European waterway network has been in the U.S.S.R., where the hauls are long, once the rivers have been regulated against strong summer floods. In 1937 the Moscow–Volga canal provided an adequate water supply and good water route for the capital. In 1952 the lower Volga–Don canal linked the Volga with ocean transport. Then, by a series of huge dams and reservoirs, already discussed, the Volga and Don were controlled, and connecting waterways and canals leading northward were improved, so that by 1965 a Volga–Baltic waterway was opened. The guaranteed navigable depth is $9\frac{1}{2}$ ft on the main rivers and 8 ft on the Kama. Vessels of up to 5,000 tons can now navigate between the White Sea, Baltic, Caspian, and Black Sea. In 1966 the Soviet river and canal systems operated 138 billion ton-miles of freight traffic, or over four times the ton-mileage of 1950. The freight movement on the Volga system alone in 1960 was estimated at 40 billion ton-miles, with cargo traffic of up to 17 million tons between Kubyshev and Volgograd. The statistics given for the U.S.S.R. in the above Table include the transport of timber, which in 1966 amounted to 69,700,000 tons and 27,600,000 ton-km.

3. Domestic and industrial uses and the problem of open-channel pollution

In most countries the domestic and industrial consumption of water increases rapidly with the growth of population, of industrialization, and of the standard of living and hygiene. Even in areas with a long-established piped supply, modern domestic appliances cause the water consumption to rise rapidly. In 1830 the British domestic user managed on less than 4 gallons (18 l.) per day, whereas by 1960 he needed over 60 gallons (270 l.) daily. In the United States each person uses on an average about 140 gallons a day, and the inhabitants of some large cities need over 180 gallons. In contrast, at Karachi the average *per capita* daily consumption is probably about 20 gallons.

Recently, owing to depletion of ground aquifers, cities have turned increasingly to rivers and overground reservoirs for an adequate water supply. The Romans built magnificent aqueducts for their Mediterranean cities. Nearly two millennia later long-distance aqueducts have become common. Probably the most striking modern scheme supplies the gold-mining towns in the Western Australian desert. Here between 1898 and 1903 a reservoir was constructed on the Helena River near Perth, and a steel pipe, mostly 30 in. in diameter, was laid from it 346 miles to Coolgardie and 351 miles to Kalgoorlie. In 1913 Los Angeles obtained a supply from a reservoir 233 miles away in the Sierra Nevada. Today this is supplemented by an aqueduct from Lake Mead on the Colorado River, a distance of 266 miles, of which 108 miles are in tunnels through mountain ranges. Recent schemes, such as that of Whyalla in South Australia (233 miles of pipeline; 1,558 ft total lift), demonstrate the distance that municipalities are prepared to go for water.

In some of these long-distance projects the water is totally lost to the source drainage basin, whereas in most smaller schemes a large proportion of the water extracted from a river is returned as sewage effluent to the same drainage basin. It is common in densely peopled areas for municipalities extracting water and discharging effluent to be spaced at intervals along the same river. This practice is safe provided the discharge of the river remains adequate and all domestic and industrial effluents are properly treated before they re-enter it.

The World Health Organization has laid down standards of desired quality for domestic water supplies. To try to improve community water supplies, which are unhygienic or unpiped and inadequate in vast areas, it co-operates with various international development associations to provide funds, surveys, and training. Thus in 1965 water surveys were made under its guidance in, for example, Ecuador and Panama, and a loan made for the development of the Johore River as a supply for Singapore.

Industrial consumption of water is usually on a bulkier scale than that of domestic users. For steam-raising and cooling purposes coal-fired electricity plants need about 600–1,000 tons of water for each ton of coal burned. The water requirement for direct cooling for a station of 2,000 MW is about 1,000 million gallons (4,500 million l.) per day, a quantity sufficient for the domestic uses of a large conurbation. The manufacture of a ton of steel requires about 250 m^3 of water, of a ton of sulphate woodpulp about 240 m^3, of a ton of soda woodpulp 320 m^3, and of a ton of woollen and worsted fabrics about 580 m^3. However, most of this water is returned into circulation and compared with irrigation, industrial and domestic users are relatively small absolute consumers of water. Yet they cause one of the greatest difficulties of water supply. Often open channels are ultimately the recipients of effluent from sewage works and industrial plants, which may result in unsafe and unsightly pollution.

In most countries waste disposal is less advanced than water supply. In Britain cholera epidemics led to a royal commission on water supply and drainage in 1843, but improved supplies led to so great an increase in the effluent from sewers and industrial concerns that by 1870 river pollution also required public investigation. Here and in many other districts of the world some rivers are discoloured and polluted sufficiently to discourage fish life and all domestic and recreational uses.

Pollution, or quality of water, is partly a question of the specific use for which the water is needed. Thus to encourage fish life, low dissolved oxygen content, and concentrations of toxic metals, ammonia and so on must be avoided; for domestic use the absence of harmful bacteria, of objectionable colouring matter, and of organic constituents with an offensive taste or smell are also important; for cooling water in thermal electric stations, the prime needs are low temperatures, relative freedom from suspended solids, and a tendency not to promote slime and corrosion. As yet very little is known about 'the chronic physiological effects' on public health of minute amounts of chemical substances contained in water supplies. The difficulty of avoiding pollution in areas where heavy artificial fertilizing and hormone spraying are practised is only just being realized.

The main remedies against excessive river pollution today are to increase the rate of discharge ('Dilution is the solution to pollution'); to increase the rate of re-aeration, by increasing turbulence by building weirs etc. and constructing purification lakes or reservoirs; to improve the flora, the clarity of the river, and cleanliness of its bed; and, inversely, to increase the purity of effluents entering the river. These ideas include a detailed monitoring system for river-water quality, whereby users can be warned immediately of the precise quality of the supply and, if necessary, undesirable water can be periodically flushed out of a drainage system. The building of purification reservoirs seems especially promising and could be combined with schemes for flushing river-beds. This latter idea, the sudden increase in flow, appeals strongly to people interested in the preservation of migratory fishes.

Open channels bring economic and human problems of a kind other than those caused by the direct consumption or use of the water. In warm climates with no resting season for hydrophytes water plants multiply and soon block shallow channels. In cooler climates the water weeds are removable by a single annual cutting, but in warm regions repeated cuttings by machine or hand are necessary. In some areas chemical sprays are becoming popular. In others, where the vegetation is suitable, herbivorous fishes, such as Chinese grass carp, puntin carp, and goramy, have been introduced. The great weed nuisance in much of the tropics is the beautiful water hyacinth (*Eichornia crassipes*), which is still spreading. In the Sudan, where it has only reached the southern watercourses, hormone spraying from boats and aircraft is among the control methods. Biological controls used locally include, first, the seacow or manatee, which is successful in parts of Guyana but is so popular with the natives as a source of meat that it is becoming scarce; and, second, certain species of snail which feed on the hyacinth.

However, these snails act as hosts for the fluke (*Bilharzia*) which causes *bilharziasis* (Schistosomiasis). At a stage in its cycle the fluke leaves the snail and moves freely in water; it then attaches itself to, and eventually enters the blood stream of, human beings that come in contact with it. A serious deterioration of the intestines and liver of the affected person then occurs. The fluke breeds in the human body before returning, as excreta, to the water snail. Bilharziasis is endemic in many parts of the tropics, and is second to malaria among the world's parasitic diseases. Unlike malaria, it is spreading, and today probably affects 150 million people. It increases in frequency near irrigation projects, especially where seasonal water supplies are replaced by perennial. In parts of Egypt its incidence rose from 5 to 75% of the adult population between 1948 and 1963 (*World Health*, World Health Organization, July–August 1964, p. 27).

The main needs to prevent bilharziasis is to control the host snails and human hygiene. The former could in some areas be at least partly achieved by lessening the shallow-water margin of watercourses and reservoirs, and also by raising and then lowering the water level to expose the snails. This latter method was successfully used in the T.V.A. scheme to control mosquitoes; marginal vegetation growth was retarded by maintaining high water during the early growing season, and at a critical time the water level was greatly lowered, so exposing the immature mosquito larvae.

The harmful inhabitants of open channels in warm regions also include carnivores ranging in size from crocodiles and alligators to the South American scourge, the small bloodthirsty piranha (*Serrasalmo*). However, these objectionable denizens are more than compensated for by the abundance of edible species.

4. Open channels and freshwater fisheries

In all countries fishing is one of the most popular recreations, and tens of millions of amateur fishermen catch a fish or two in their leisure hours. Professional and part-time freshwater fishermen and fish-farmers are commonest in monsoon Asia, where fish is an indispensable part of the diet and fish sauces, such as the *prahoc* of Cambodia, flavour almost every meal.

Of the world's fish catch in 1966, freshwater and diadromous (migratory) species provided about 7,910,000 metric tons or 13·9%. Of this the diadromous fishes supplied 1,760,000 tons, the main species being salmons, trouts, smelts, capelin, and other *Salmonidae* (1,180,000 tons); shads, milkfish, and similar species (520,000 tons); freshwater eels (41,000 tons); and sturgeon (about 16,000 tons). The surprising growth in the catch of capelin in 1966 (521,000 tons, of which Norway caught 379,000 tons and Iceland 124,900 tons) should not be allowed to hide the long-standing importance of salmons. Among anadromous species (which migrate up rivers to spawn) the salmons are supreme. In a single year at sea they can put on up to 10 lb in weight. They are caught in rivers and in tidal estuaries, and their importance can be judged from the catches on the Pacific coasts and rivers in 1966, when over 260,000 metric tons were landed in and off the United States and Canada, and about 190,000 tons in and off Japan and the eastern U.S.S.R. In that year the Scottish catch was 1,300 tons.

These diadromous fishes depend for their life cycle on a freshwater channel with a depth and discharge of healthy water sufficient to enable them to reach their spawning grounds. They happen to frequent the coastal rivers of glaciated areas, where hydroelectric dams are most common. The obstruction to migration upstream caused by modern dams is insurmountable, except perhaps for eels (which migrate to the sea to spawn), while the injuries or fatalities caused by turbines to small fry migrating downstream may be over 50% of those passing through. Consequently, there has been a constant disagreement between fish conservationists and water-resource developers, and today elaborate precautions are taken to allow fish migrations to continue unharmed. The normal practice is to build fish ladders or other forms of stepped waterfalls which the migratory fish use to by-pass the dam. Sometimes these devices prove relatively ineffective and waste too much water. Other methods involve the stripping of spawn and the artificial reproduction of fry in hatcheries and the restocking or artificial seeding by collected spawn of the upstream spawning grounds. These, and a better river flow in spring to encourage the fish, are the methods being used, for example, on the Volga system, where the huge dams have disturbed the conditions for semi-migratory roach, bream, etc., and obstructed the route to the spawning grounds for migratory sturgeon, whitefish, and some forms of herring. The U.S.S.R. (22,000 tons in 1962; 15,100 tons in 1966) produces nearly all the world's commercial catch of sturgeon, a large fish which yields a fine flesh as well as caviare and isinglass. In 1932–6 about 75% of the sturgeon catch came from fisheries on the Volga and north Caspian, but yields here declined markedly in the 1960s. On the other hand, the vast new reservoirs above the Volga–Don dams are becoming important fishing grounds for non-migratory species, and nearly one-fifth of Soviet Russia's entire fish production (5,350,000 tons in 1966) comes from fish farms and inland waterways.

The advantage of manipulating reservoir releases so that a surge of fresh water encourages and enables anadromous fish to migrate from estuaries to their spawning grounds has also been demonstrated on the Roanoke River in North Carolina. Here striped bass apparently responded to such surges, which also brought the additional benefits of improving the stream bed for fish food organisms. The reservoir release, however, may be a loss to hydroelectric production, as only a volume sufficient to reach the estuary is likely to affect the maximum number of fish.

Caution may be necessary in some localities after sunny weather has stratified the water in deep reservoirs, causing the bottom layers to become low in oxygen content. This bottom water, if passed through turbines to the river downstream, may be harmful to fish life there. Such a possibility can be avoided by selective withdrawal of the better-oxygenated surface layers of the reservoir and by artificial aeration below the dams.

No such difficulties occur in connection with fish that spawn without migrating appreciable distances. These non-migratory freshwater species, which yielded a total catch of 6,150,000 metric tons in 1966, are ideally suited to lakes, reservoirs, and pisciculture generally. Provided they have adequate food, and water with an oxygen saturation above about 75% and reasonably free of excreta, these fish will thrive in still or almost still water. The protein-producing potential of rivers and freshwater ponds is enormous, especially in the tropics. A suitable reservoir stocked with carp will produce 200–400 lb of fish per acre per year in central Europe; 500–600 lb in Alabama; and up to 1,000 lb in parts of Monsoon Asia. In Java the milkfish (*Chanos chanos*) and other species reach marketable size in less than one year; here and elsewhere in south-eastern Asia spawn or fry are often bought by the thousand for stocking tanks, where they will grow to 5 or 6 lb in nine months. River fisheries and pisciculture, such as keeping carp on paddy-fields, play a large part in the economy of monsoon countries. In India 447,500 tons or 36% of the total national fish catch in 1966 came from rivers and freshwater reservoirs. Then the annual freshwater catch of Indonesia was 281,400 tons (28% national total catch); of Pakistan 231,800 tons (56% total); and of Cambodia 125,000 tons (76% total). In pisciculture there are fascinating specialities: for example, in 1966 Japan produced mainly by aquiculture 19,800 metric tons of eels (48% world total) and Denmark 15,100 tons of trout (52% world total commercial production). In addition, where malaria is prevalent fish play a vital role in keeping down mosquitoes, and no stagnant or slow-moving water should be without edible species. Buffalo fish are useful denizens of the ricefields of the Mississippi.

5. Open channels and the preservation of wild life

The preservation of biological species, including rare species of fish, concerns the planner of water resources. This is often a question of prohibiting or removing industrial pollution and of protection also against other human interference and predators. The reintroduction of locally extinct or failing species often brings satisfactory results once the river habitat has been restored, if necessary, to its original quality. Multiple-purpose basin planning usually incorporates the provision of lakes which act incidentally as resting places for migratory birds such as geese. In areas where the streams dry up seasonally river life can be greatly increased by the building of dams, which may provide a small dry-season flow or at least provide a refuge during the drought. This has been done successfully on some of the small trout streams of the Sierra Nevada in California.

6. Social uses of open channels

In a few countries some springheads and rivers acquire great religious significance. The ice cave near Gangotri (10,300 ft above sea-level), which is the source of the Bhagirati, and the junction of this headstream with the other main headstream of the Ganges at Devaprayag are famous pilgrimage places. The chief pilgrimage centre, however, is lower down at Allahabad, where the Jumna joins the mature Ganges. Here the bathing festival of Magh Mela is usually attended by 250,000 Hindus, and every twelfth year the special festival of Kumbh Mela is attended by 1 million Hindus, who wash away their sins in the sacred river.

In nearly all countries rivers and lakes are important directly for recreation in addition to fishing for sport and subsistence. Bathing, swimming, and washing, personal and sartorial, are universal uses in warm lands. In societies where leisure is common and fresh air a cult or a necessity the construction of dams and reservoirs often makes recreation perennial rather than seasonal, or at least lengthens its season and renders it safer and easier than on unregulated rivers. Reservoirs become the scene of boating, sailing, water ski-ing and, where free of pollution, swimming, while their shores provide camping grounds for tourists. In some countries, especially the United States, recreation is an important component of many water projects, and in most industrialized societies the appreciation of the value of recreation as a source of pleasure and income has greatly increased. Fifty years ago the Hoover Dam (Lake Mead) project catered on a large scale for boating, sailing, and swimming in the hot sunny climate. In 1933 the Muskingum Watershed Conservancy multiple-purpose scheme covering about 8,038 square miles of territory included fourteen reservoirs, which provided 16,000 acres of lake and 365 miles of shoreline. Each lake or reservoir was encircled by a landscaped strip at least 100 ft wide perpetually reserved for public use. Within fifteen years about $2\frac{1}{2}$ million visitors were coming annually. Recently a scheme for a 37-mile long reservoir on the Delaware River above Tocks Island allowed about one-third of the total estimated cost for purposes of recreation and provided for a capacity of over 10 million visitor-days annually.

Many of the great dams themselves become noted tourist attractions, for example, the Glen Canyon dam with its tourist viewpoints and the Aschach dam on the Danube in Austria.

REFERENCES

GENERAL

UNITED STATES GOVERNMENT [1950], *A Water Policy for the American People*, 3 vols., (Washington, D.C.).

FLOTABILITY AND NAVIGATION

Atlas of Finland [1960] (Helsinki). (See plate 24.)

Atlas of Sweden [1953–68] (Stockholm).

CANADIAN GOVERNMENT [1967], *Traffic Report of the St. Lawrence Seaway, 1966* (Queen's Printer, Ottawa).

MEAD, W. R. [1958], *An Economic Geography of the Scandinavian States and Finland* (Univ. of London Press), 302 p.

MILLWARD, R. [1964], *Scandinavian Lands* (Macmillan, London), 488 p.

ROM, V. Y. [1961], The Volga–Baltic waterway; *Soviet Geography*, 2 (9), 32–43.

TAAFE, R. N. [1964], Volga River transportation; In Thornan, R. S. and Patton, D. J., Editors, *Focus on Geographic Activity* (McGraw-Hill, New York), pp. 185–93.

UNITED NATIONS [1967], *Annual Bulletin of Transport Statistics for Europe* (New York).

UNITED STATES [1967], *Annual Statistical Abstract* (Washington, D.C.).

VENDROV, S. L. *et al.* [1964], The problem of transformation and utilization of the water resources of the Volga River and the Caspian Sea; *Soviet Geography*, 5 (7), 23–34.

DOMESTIC AND INDUSTRIAL WATER USES AND POLLUTION

ISAAC, P. C. G., editor [1967], *River Management* (Maclaren, London), 258 p.
See particularly the following chapters:

BRIGGS, R. *et al.*, The monitoring of water quality, pp. 38–55.

HOUGHTON, G. U., River-water quality criteria in relation to waterworks requirements, pp. 153–67.

LESTER, W. F., Management of river water quality, pp. 178–92.

LOVETT, M., Control of river quality, pp. 193–8.

MERCER, D., The effects of abstractions and discharges on river-water quality, pp. 168–77.

WOLMAN, A., editor [1962], *Water Resources;* National Academy of Science, Publication 1000-B, National Research Council (Washington, D.C.).

WORLD HEALTH ORGANIZATION [1963], *International Standards for Drinking Water*; 2nd edn. (Geneva).

OPEN CHANNEL AND FRESHWATER FISHERIES

FOOD AND AGRICULTURE ORGANIZATION OF THE UNITED NATIONS [1967], *Yearbook of Fishery Statistics*, 1966, Vol. 22 (Rome).

ISAAC, P. C. G., editor [1967], *River Management* (Maclaren, London), 258 p.
See particularly the chapters by:

BRAYSHAW, J. D., The effects of river discharge on inland fisheries, pp. 102–18.

HULL, C. H. J., River regulation, pp. 86–101.

7.III(ii). Rivers as Political Boundaries

ROBERT P. BECKINSALE
School of Geography, Oxford University

1. Rivers as internal administrative boundaries

Rivers are commonly used as administrative boundaries within a state. Streams, being occasionally impassable or awkward to cross, form recognizable limits for minor administrative units such as parishes in England, many of which date back to the eighth and ninth centuries A.D. Yet in England the larger administrative units, such as counties or shires, are rarely bounded by rivers, largely because they were conceived as being centred upon a defensive town situated on a water-way. The main exception is the Thames, which acted as a defensive barrier between the kingdoms of Wessex and Midland Mercia and which is today a county boundary for most of its length. When a river acting as a boundary changes its course, naturally or with artificial aid, the original boundary line is retained and becomes of considerable geomorphic interest, as can be seen from the lower Dee and Dove (fig. 7. III(ii).1).

Similar domestic river boundaries are more common in states or provinces colonized in recent times by Western powers. In Australia the Murray between New South Wales and Victoria, and in Canada the Ottawa are inter-provincial boundaries. But the use of internal river boundaries reaches its maximum in the United States, where, for example, the Delaware, Potomac, and Savannah east of the Appalachians and the Ohio and Mississippi west of the Appalachians are state boundaries for most of their length. This method of using ready-made river lines would no doubt have caused endless interstate friction had not the Federal Government financed all improvements for navigation.

2. Rivers as international boundaries

In monsoon Asia the national territories of paddy-oriented nations tend to extend across floodplains and rivers, and the only notable river boundary is the middle Mekong (Thailand–Laos), here rapid and gorge-like. Even in arid Asia national territories usually include both sides of a river. Thus in the U.S.S.R. the lower Amu-Dar'ya (ancient Oxus) is avoided as a boundary except for a short stretch, although its steeply incised upper course acts generally as a boundary between Afghanistan and Tadzhikistan-Badakshan. In fact, the only great international river boundary in Asia is the Amur and Argun rivers between Manchuria and the U.S.S.R.

In Europe the Rhine was a defence line in Roman times, but international

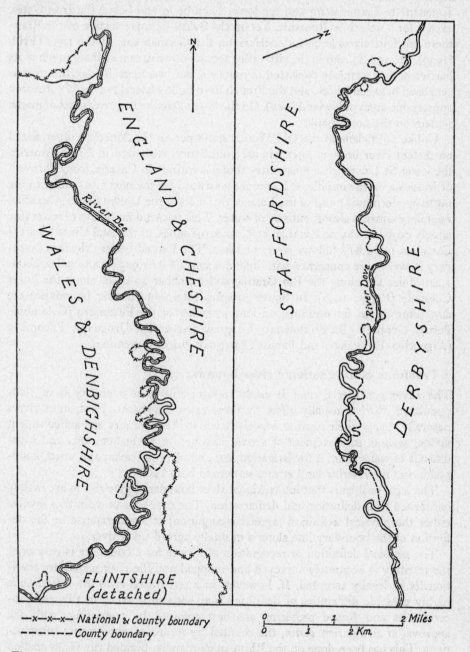

Fig. 7.III(ii).1 Administrative boundaries as evidence of changes in a river's course: the lower Dee and lower Dove, England. Embankments or levees are not shown.

river boundaries are uncommon. Apart from short stretches such as the Torne River (Sweden–Finland), the only notable examples are the middle Rhine from Konstanz to Lauterbourg and the lower Danube in and below the Iron Gates, except for a stretch in Romania. Yet on the Swiss–German section of the Rhine there are four sizeable Swiss enclaves on the German side of the river (Früh [1939], pp. 494–8), and in the rift-valley section downstream of Basle much of its barrier nature formerly consisted of marshes and braidings. Today the river is canalized between levées, and the French have built a lateral canal and cultivated appreciable areas of lowland (*ried*). On the lower Danube large expanses of marsh remain on the north bank.

Unlike politicians in the Old World, statesmen in the Americas often seized on distant river lines as international boundaries. Although in North America the lower St Lawrence, a great entry route, is entirely in Canada, from Cornwall or Massena on the middle St Lawrence westward for the next 1,715 miles to the north-westernmost point of the Lake of the Woods, the United States–Canadian boundary consists almost entirely of water. This tremendous length of water line largely explains why no less than 55%, or 2,198 miles, of the total United States–Canadian boundary follows rivers or lakes. The United States–Mexican boundary is even more concerned with inland water. Of its 1,905 statute miles, about 1,210 miles are along the Rio Grande and a further 20 miles along the lower Colorado (Boggs, 1940). In South America international river boundaries are also common, as for example, on long stretches of the Putumayo (Colombia–Peru); Guaporé (Brazil–Bolivia); Uruguay (Argentina–Uruguay); Pilcomayo (Argentina–Paraguay); and Parana (Paraguay–Brazil–Argentina).

3. Problems of international river boundaries

The advantage that a river is easily recognizable as a boundary is in well-populated districts usually offset by three main drawbacks. First, most rivers naturally change their courses, especially where they meander and suffer violent floods; second, improvement of a river channel for navigation, etc., and withdrawals of water from it for irrigation, etc., affect both banks; and third, flood-plains and flat riverine land attract settlers to both banks.

The legal problems that international river boundaries give rise to are mainly concerned with definition and demarcation. The problems of definition involve either the physical definition (agreed recognition) of a watercourse or the definition of the boundary line along a mutually agreed-upon river.

The physical definition or recognition of a river for a boundary is easy once the territory is accurately surveyed and mapped and the river names are traditionally or legally accepted. If, however, in a recognized river the waterline is highly unstable difficulties of definition may constantly arise. In this case uncertainties and future problems are best resolved by restricting, with the approval of all riparian states, the channel by means of engineering constructions. This has been done on the Rhine in its marshy, braided rift-valley section and on the Rio Grande. Near the latter when the United States–Mexican treaties were signed in 1848 and 1853 there were few settlers and little economic interest

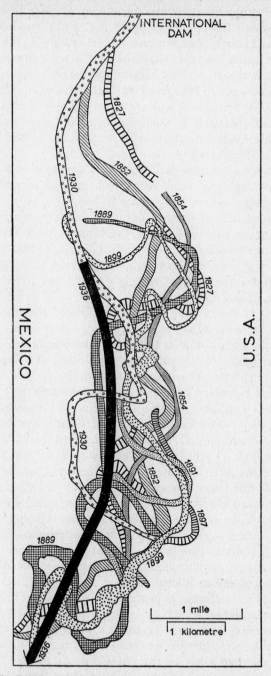

Fig. 7.III(ii).2 The Rio Grande near El Paso–Ciudad Juarez, showing some of the main shifts in the major channel from 1827 to 1936 (Adapted from Boggs, 1940).

except in navigation. The settlement and variety of water uses increased rapidly, but the channel remained highly unstable due to excessive silting and frequent cut-offs of meanders during violent floods (fig. 7.III(ii).2). Between 1907 and 1933 the river bed near El Paso was raised 8 ft by silting. In the following years, under an International Boundary Commission with wide powers, a flood-control reservoir was built at Caballo, New Mexico, and the river bed straightened (from 155 to 86 miles between El Paso and Fort Quitman) and endyked so that silting virtually ceased. The parcels of land cut off by the new international boundary along the rectified channel were exchanged in equal proportions of about 3,500 acres on each side. On the basis of benefits received, the United States paid 88% of the total costs (Boggs, 1940).

In the above cases the river itself was not in doubt, whereas in partially explored areas the river may need definition or mutual recognition before demarcation can begin. There exists today a dispute about a frontier between Guyana and Surinam, two former European colonies, which involves 5,800 square miles in the triangle between the Courantyne River (named as the boundary) and the so-called New River. Guyana claims that the New River is really a tributary of the Courantyne, whereas Surinam claims that it is in fact a continuation of the Courantyne wrongly named New River on the original European maps (*The Times*, 1968).

The problem of determining which river formed the master stream of a drainage basin was argued fully in the Argentine–Chile Frontier controversy of 1966 in which the author acted as hydrological adviser to the Republic of Chile (Foreign Office, 1966). The dispute originated largely in a Boundary Award of 1902 when the territory involved was largely unexplored. This Award stated that the international boundary ran from a fixed boundary post 16 up the Rio Encuentro from its junction with the Rio Palena (or Carrenleufú) to its source on the western slopes of the Cerro de la Virgen. Later it was discovered that the river joining the Palena opposite boundary post 16 rose at a considerable distance from the Cerro de la Virgen. In the recent arbitration it was accepted that the lower course of the river opposite boundary post 16 was the Rio Encuentro. However, above this lower course, which is partly in a deep gorge, the river consists of two main headstreams: one, hereinafter called A, which rises in a cirque high up on the mountain flanks to the east and approaches the main valley-floor at a sharp angle; and another, hereinafter called B, which rises at a lower altitude and drains mainly the northern part of a broad glaciated valley that continues the general direction of the lower gorge. As the tributaries and their sources lacked any traditional names, the arbitrators had to decide which tributary was the main channel or continuation of the lower Encuentro on other historical evidence and on scientific grounds. Scanty historical evidence showed that a few observers considered channel A was the main channel. Modern scientific hydraulic principles left no doubt that A was the master stream. Consequently, the arbitrators decided that channel A was the Rio Encuentro.

The scientific arguments used by Chile to show that channel A was the master stream were that, compared with channel B, it had:

1. greater length (exact measurements were given in all the following linear, areal, and volumetric measurements);
2. greater discharge (based partly on gauge measurements and partly on scientific precipitation/altitude relationships);
3. greater drainage basin;
4. greater geomorphic age and incision (being in existence in its upper course when stagnant ice still occupied the valley floor now drained by channel B);
5. at and near their junction, channel B had a greater gradient of bed and of water level than channel A;
6. strong similarities of bed load existed only between channel A and the lower course below the junction of channel A and channel B.

The arbitrators considered that items 1–3 were the principal criteria to be applied in a problem of this kind and that items 4–6 did not contribute significantly to its solution.

Argentina laid much stress on the lineal continuity of channel B with the lower trunk river, and the Court admitted that this could be an important factor and quoted Strahler's use of it as equal to 'longest total stream length as a factor, when continuing a single-channel profile headward into channels of lower order (Strahler, 1964). However, the Court recognized Chile's contention that the lineal continuity referred to was primarily continuity of valley-form (of the glaciated trough), whereas more importance should be attached to 'the continuity of the general force of the river', which was more evident in channel A.

Argentina introduced the Strahler method of stream order-numbering to show that at their junction the two channels were of the same order, whereas Chile suggested that on the Horton system channel A was the master channel. However, the Court did 'not consider that the two different methods of order designation (applied as they are to maps of different degrees of accuracy and on different scales)' helped to resolve the problem. However, the author retains the opinion that Horton's method of 1935 and 1945 (Leopold, Wolman, and Miller [1964], pp. 134–5) would normally provide a satisfactory solution to such a problem, as it does select and extend to its headsource the chief order stream in a drainage basin, whereas Strahler's 1952 simplification of Horton's method was deliberately designed to avoid making that decision.

In its conclusions the Arbitration Court ordered part of the new boundary to 'follow the thalweg of the Encuentro', although fieldwork had shown that in this mountainous drainage basin there is in fact usually no thalweg in the strict sense, as most of the river channels have almost flat beds lined irregularly with large pebbles and boulders. As, however, will now be shown, it is usually very difficult to define and demarcate precisely a river boundary.

In water, especially where flowing, marks are often impracticable to fix or to maintain, and are often thought unnecessary, as they are liable to be a danger or nuisance to both parties when the river or lake is used for certain lawful purposes, such as navigation and fishing. Water boundaries are usually demarcated by reference to land marks, and are often in practice marked in detail only on the official reference maps. The demarcation, however, should be based on a precise definition. Such boundaries follow either

1. the shore;
2. the median line;
3. the thalweg; or
4. some arbitrarily selected line.

The fourth type occurs today, for example, as a parallel of latitude on Lake Victoria between Tanzania and Uganda and as azimuths or straight-line courses on the United States–Canadian water boundaries.

The first type is rare, and probably the chief existing example is on the Shatt al Arab, where in 1914 the boundary was drawn along the low-water line of the left or Persian bank, thus depriving Persia of free access to the river fairway. Later, Iraq succeeded to Turkish rights, and there were many incidents between Persian and Iraqi officials on the Shatt al Arab before a Treaty of Friendship was signed. The agreement provided for the inclusion of anchorages in the Shatt al Arab off both Abadan and Khorramshahr in Persia and the general retention of the rest of the 1914 frontier.

The third type, the thalweg or deepest channel, was commonly used in international agreements when navigation interests predominated. The definition here refers to the cross-section of a river's bed, although in some early treaties thalweg and the 'middle of the main channel' were assumed to be coincident. The concept envisages an 'uninterrupted line determined by the deepest places in the bed' (Kaeckenbeeck [1918], p. 176; Haataja [1927]), and although the thalweg is not completely stable, it offers 'greater stability than the middle line of a stream'.

The second type of definition or demarcation, the middle line or median line, could entail great inconvenience, as it changes its position with changes in the water level and in the shape of the bed (e.g. the Rhine between Germany and Switzerland). Moreover, the deep-water channel (thalweg) might be entirely on one side of the middle line, so depriving one riparian state of beneficial possession. Unfortunately the median line failed to develop precise concepts. It was variously defined as:

1. the 'middle' of a watercourse or a line at all points equidistant from each bank;
2. a line paralleling the general line of the banks and dividing the horizontal surface area of the water into two equal parts;
3. in the case of a lake, a line along the mid-channel dividing the navigable portion and being at all points equidistant from shoal water on each shore (*International Waterway Comm. compiled reports, p. 578*).

It was not, however, until 1930 that the following exact definition was proposed by Boggs [1940, pp. 181–2]. 'A median line is a line every point of which is equidistant from the nearest point or points on opposite shores of a lake or river' (see also Boggs [1937] and Burpee [1938]). Ambiguities were now virtually impossible, although islands may still cause difficulties.

The United States–Canadian water boundaries exemplify the increasing precision in definition and demarcation. By the treaty of 1783 (Art. 2) the boundary followed the 'middle' of twelve lakes, rivers, and inland straits. However, curved lines proved impracticable, and in 1908 the International Commission was empowered to replace them by 'a series of connecting straight lines'. The boundary from Cornwall on the St Lawrence to the mouth of the Pigeon River on Lake Superior now consists of 270 straight-line courses (although the two of these that follow parallels of latitude mathematically speaking are not *straight*; that is are not *azimuths*). Smaller, deeply indented lakes and smaller winding rivers are more difficult to demarcate. Thus, the international boundary between the mouth of the Pigeon River and the north-westernmost point of the Lake of the Woods now consists of 1,796 straight-line courses. Farther east the boundary in the 'centre of the main channel or thalweg' of the St Croix river now comprises 1,008 straight-line courses, and that along Hull's stream consists of 766 straight-line courses in a distance of 26·6 miles. No wonder Boggs [1940, p. 54] calls this 'one of the best-marked frontiers in the world'.

4. International boundaries across river channels

Most great and many smaller rivers are international, although international boundaries commonly follow watersheds. The boundary that cuts most ruthlessly across rivers and drainage basins is that between Canada and the United States from Lake of the Woods to the Pacific which follows parallel 49° N for about 1,257 miles. In this section waterway problems have been almost continuous and have been successfully settled by an International Joint Commission. A similar commission on the United States–Mexican boundary on the Rio Grande was equally successful, and under a treaty of 1945 was given wider powers of collaboration 'in order to obtain the most complete and satisfactory utilisation' of the water available.

Examples of integrated water-resource schemes elsewhere are becoming less rare. The Nile Waters Agreement has resulted in, for example, the Owen Falls Dam at the outlet of Lake Victoria, the Sadd-el-Aali barrage and, indirectly, in several Sudanese schemes. Here the working arrangements are implemented by the irrigation departments of Egypt (U.A.R.) and of the Sudan Republic.

A United Nations' *Report* [1958] stresses 'the inadequacy of international law' in respect to multiple-purpose, integrated-basin development schemes and sets out the principles drafted by the International Law Association at Dubrovnik in August 1956 as being likely 'to aid adjustment and agreement'. Since these principles were suggested at least two notable international drainage-basin development agreements have been successfully negotiated.

The Indus Water Treaty of 1960 originated largely because the north-eastern part of the Indus basin was sub-divided politically in 1947, leaving the upper courses in India and the lower courses of the chief tributaries mainly in Pakistan (United Nations [1966], pp. 47–66). At the time of partition, of the annual flow of the Indus system (about 207,500 million m³ or 168 million acre-feet) about 40% ran out to sea at the delta, 16% was lost through seepage, 39% used in irrigation canals in Pakistan, and 5% for similar purposes in India. The latter diverted water mainly from the three eastern rivers (Ravi, Sutlej, and its right bank tributary the Beas), which have a total annual discharge of 40,350 million m³. But Pakistan was already using 14,800 million m³ from these eastern rivers, whereas India, to meet all her local projects, needed all told about 38,000 million m³. On the other hand, the three western rivers, Chenab, Jhelum, and Indus, have four times the discharge of the eastern rivers and can be easily used in the plains only by Pakistan.

After thirteen years of negotiations, greatly aided by the World Bank, the Treaty agreed that the water of the three eastern rivers shall be for unrestricted use by India and that of the three western rivers for unrestricted use by Pakistan. Thus Pakistan will in the next ten or thirteen years make good the 14,800 million m³ it withdrew from the eastern rivers by extra withdrawals from those in the west. In the meanwhile India will supply to Pakistan from the eastern rivers the deficit between 14,800 million m³ and the extra amounts withdrawn by Pakistan from the western rivers. From these western rivers Pakistan will receive about 167,000 million m³ minus a relatively small proportion which will be diverted to India for various purposes. India may develop hydroelectric stations (to some specific design criteria) and build storage reservoirs of about 4,500 million m³ total capacity on the headstreams of the western rivers in India for general purposes, particularly flood control. The Treaty is implemented by a Permanent Indus Commission and contains mutual obligations on exchanging hydrologic data, etc., operating the river work, maintaining river channels, floating timber, pollution and procedures about disputes. It provides 'an admirable instance of international co-operation' (United Nations [1966], p. 66).

Its counterpart in the New World is the Columbia River Treaty of 1964 between the United States and Canada. This 'major milestone in the history of international river development in North America' (Sewell, 1966) required twenty years of studies and negotiations. The instigation came partly from proposals to build in the United States section of the Columbia dams which would back up water into Canada (cf. the Sadd el Aali dam and the Sudan). The opportunity for integrated basin development led in 1944 onwards to studies of the whole Columbia River basin in the United States and Canada. Emphasis, however, was put on hydroelectric generation and on flood control (fig. 7.III(ii).3). The Report of 1959 produced three alternate schemes of maximum development, each providing about 17 million kW of firm power and about 50 million acre feet of water storage at a total cost of about $4 billion. As about 23 million acre feet of this storage would be developed in Canada, where the head or fall was relatively small, its storage value would lie mainly in supplies to power-stations downstream in the United States, where also flood-control benefits would be greatest. Consequently, it seemed that some inducement or compensation should be given to Canada to develop these storage units. This concept of 'downstream benefit sharing' caused difficulties and led to a narrowing

Fig. 7.III(ii).3 The main engineering projects on the Columbia River drainage system (From Sewell, 1966).

Only the larger natural lakes are shown. The ultimate installed hydroelectric capacity of some of the larger stations is: Grand Coulee 5,574,000 kW; John Day 2,700,000 kW; Mica 2,000,000 kW; The Dalles 1,743,000 kW; Chief Joseph 1,728,000 kW; Wanapum 1,330,000 kW; Priest Rapids 1,262,000 kW; McNary 986,000 kW.

in the treatment of the integrated river-basin problem. According to Sewell [1966, p. 150], the negotiators aimed at reaching agreement and at making a benefit/cost profit rather than producing an overall scheme that would yield the greatest net benefits. Some of the difficulties arose because the United States' stretch of the river was already partly developed and their negotiators wished for an unfinished project in an advanced stage of design to be included in the international agreement. On the Canadian side there were difficulties due, for example, to Province–Federal politics and to vague alternative sources of hydroelectric power.

Eventually the Treaty was signed by Canada in 1964. Under its provisions 15·5 million acre feet of storage will be built on the headwaters of the Columbia River in Canada at three sites: Arrow Lakes (7·1 million live storage), Duncan Lake (6·4 million), and Mica Creek (7 million, rising later to 12 million live storage). This storage will be used mainly for increasing the winter flow of the Columbia and in reducing flood damage. The benefits will be shared by the two states. In 1964 the power benefits were estimated at 2·6 million kW capacity and 13 billion kWh output, but these have since been reassessed at 2·75 million kW and 15 million kWh. The flood-control benefits were assessed at $126 million. Canada elected to sell her share of the increased power to the United States for the next thirty years for a lump sum of $253·9 million (U.S.) ($273·3 million Canadian), which was enough to pay for the construction of the three storage dams and a considerable proportion of the cost of installing generating units at the Mica project (dam height 645 ft; ultimate installed capacity 2 million kW, for use exclusively in Canada). Canada was also paid $64·4 million (U.S.) as her share of flood-control benefits. The United States had the option to build the Libby dam on the Kootenay River in Montana, the reservoir of which will stretch several miles into Canada (fig. 7.III(ii).3). Either country is allowed to take water for domestic uses at any time and Canada to divert up to 1·5 million acre feet from the Kootenay to the Columbia for power purposes after twenty years. No doubt, as the parties hoped, the Treaty will stand out as 'an example of large-scale international co-operation, hopefully to be imitated elsewhere'. However, the negotiations showed that a truly comprehensive approach is difficult to put into practice and, in the opinion of some, that a higher proportion of the investigation costs should have been spent on economic, as distinct from engineering, studies.

REFERENCES

ADAMI, V. [1927], *National Frontiers in Relation to International Law*; Translated by Behrens, T. T. (Oxford University Press, London). (See especially pp. 199–210 and 218–19.)

BOGGS, S. W. [1937], Problems of water-boundary definition; *Geographical Review*, 27, 445–56.

BOGGS, S. W. [1940], *International Boundaries* (Columbia University Press, New York), 272 p.

BURPEE, L. J. [1938], From sea to sea; *Canadian Geographical Journal*, 16, 3–32.

DEPARTMENT OF EXTERNAL AFFAIRS, CANADA [1964], *The Columbia River Treaty: Protocol and Related Documents* (Ottawa).

FAUCHILLE, P. [1925], *Traité de droit international public;* Vol. I, Part 2 (Paris).

FOREIGN OFFICE [1966], *Award . . . for the Arbitration of a Controversy between the Argentine Republic and the Republic of Chile;* Reference S. O. Code No. 59–163 (H.M.S.O., London).

FRÜH, J. [1939], *Géographie de la Suisse;* Vol. 2 (Librarie Payot, Paris).

GLOS, E. [1961], *International Rivers: A Policy-Oriented Perspective* (Singapore).

GRIFFIN, W. L. [1959], The use of international drainage basins under customary international law; *American Journal of International Law,* 53, 50–80.

HAATAJA, K. [1927], Questions juridiques surgies lors de la révision de la frontière finlandaise entre le golfe de Bothnie et l'océan Glacial; *Fennia,* 49, 1–46.

INTERNATIONAL COLUMBIA RIVER ENGINEERING BOARD [1959], *Water Resources of the Columbia River Basin;* (Ottawa).

INTERNATIONAL JOINT COMMISSION [1959], *Report on Principles for Determining and Apportioning Benefits* . . . (Ottawa).

JONES, S. B. [1945], *Boundary Making* (Washington).

KAECKENBEECK, G. [1918], *International Rivers* (London). (See especially p. 176.)

LAPRADELLE, P. DE [1928], *La frontière: étude de droit international;* Vol. 1, Part 2 (Paris).

LEOPOLD, L. B., WOLMAN, M. G. and MILLER, J. P. [1964], *Fluvial Processes in Geomorphology* (Freeman, San Francisco and London), 522 p.

PRESCOTT, J. R. V. [1965], *The Geography of Frontiers and Boundaries;* (Hutchinson's University Library, London).

SEWELL, W. R. D. [1966], The Columbia River Treaty: Some lessons and implications; *The Canadian Geographer,* 10, 145–56.

SMITH, H. A. [1931], *The Economic Uses of International Rivers* (P. S. King and Son, London).

STRAHLER, A. N. [1964], Quantitative geomorphology of drainage basins and channel networks; In Chow, V. T., Editor, *Handbook of Applied Hydrology,* Section 4-II.

The Times [1968], News Item, 2 February, p. 6 (London).

UNITED NATIONS [1958], *Integrated River Basin Development* (New York).

UNITED NATIONS [1966], *A Compendium of Major International Rivers in the ECAFE Region;* Water Resources Series No. 29 (New York).

8.III. The Economic and Social Implications of Snow and Ice

J. ROONEY

Department of Geography, Southern Illinois University

Snow represents both a valuable resource and a menacing natural hazard. Although its economic utility is difficult to evaluate precisely, we can make some meaningful estimates of its contribution to agriculture, recreation, domestic and industrial water supply, and even to our less-tangible aesthetic needs. For example, most of the irrigated agriculture in the western United States, notably in the Central Valley of California, is supported by meltwater from snowpack; while in the past twenty years the number of Americans who ski has risen from 50,000 to approaching 4 million, representing only one aspect of the rapidly increasing economic benefits associated with snow-oriented recreation.

Knowledge concerning the equally significant negative social and economic impact of snow and ice is probably more sketchy. It is believed that the costs of snow-caused disruption and loss of life, plus the funds spent to combat the hazard, amount to at least 1 billion dollars a year in the United States.[1] During the period 1958–66 the U.S. Weather Bureau directly attributed more than 500 deaths to severe snowstorms, and many more persons died in traffic accidents resulting from more minor storms or prematurely from some form of overexertion, such as snow clearing.

1. Snow as a water-supply source

Snow is the principal source of water not only in many of the world's mountainous areas but also in the densely settled plains and valleys adjacent to them. It is estimated that about one-half of the streamflow in the western United States is of snowpack origin. Considerable study has been devoted to snow-zone management in the Sierra Nevada, and it is believed that California is representative of many other areas characterized by marked contrasts in elevation and vegetation. Approximately 51% of California's streamflow originates in the

---

[1] According to the U.S. Weather Bureau, damage attributed to the snow hazard in the United States ranged from an absolute minimum of $1,502,550 in 1964 to a maximum of $738,841,500 in 1958. However, these estimates are based on the high and low estimates of the U.S. Weather Bureau for those storms which were deemed severe enough for publication in *Storm Data*, U.S. Weather Bureau, Department of Commerce, Washington, D.C. They probably represent no more than 20% of the damage attributable to snow and ice each year in the United States.

snowpack zone, 32% comes from the lower forest zone, and the remainder from the foothills and lowland areas.

One of the primary aims of any water-management strategy is to reduce the annual variability of supply, and it is here that the snow zone takes on particular significance. Not only is the yield from the snow zone the most important source of water, but in the most critical dry years it is far more dependable than sources in other zones. For example, in the southern Sierra Nevada less than one year in ten is a dry year (defined as having less than one-half the mean annual precipitation) in the snow zone, whereas two to three are dry in the coniferous forest zone, and more than four in ten are dry in the foothill zone. The variability of water supply from the snow zone is further reduced by the low evaporation from the snowpack.

The economic impact of snow-zone water in both California and the whole of the western United States has consequently been tremendous. Cash-farm income in California ranks first in the United States, and nearly all crops are irrigated. A comparison of the flourishing cotton, vegetable, fruit, beet, grain, and potato agriculture of the irrigated sections of the San Joaquin Valley with the poor grazing lands of its dry western foothills is all that is necessary for one to appreciate the value of the snowpack as a water-supply source.

2. Recreation

Participation in recreation associated with snow and ice has been expanding faster than most other forms of recreational activity, with ski-ing the most prominent beneficiary. In the United States 'ski-ing activity days' have been increasing at the rate of 15% per year, and there has been an annual increase of 20% in the ski-ing population. According to the Ski Trade Association, almost 3 million 'serious' skiers spent approximately $1 billion in connection with the sport during 1967–8 alone, and over 100,000 winter-sports enthusiasts were flown to Denver by a single airline in that year.

To accommodate this ballooning United States' demand nearly 1,200 ski resorts are now in operation, over half of them equipped with chair lifts. Most of the spectacular ski areas are located in the sparsely settled western section of the country, necessitating extensive travel and adding to the general economic significance of the ski industry (figure 8.III.1). Another economic multiplier effect has been the development of sophisticated residential estates in association with ski resorts, such as the flourishing complexes at Jackson, Wyoming, and Snow Mass at Aspen, Colorado, the latter involving a planned $75 million investment.

It should not be thought that all the economically important recreational aspects of snow are limited to ski-ing. Snowmobiling is another activity which has experienced a phenomenal expansion during the past five years. Snowmobile production in North America has grown from 15,000 units in 1963 to 175,000 in 1967, with many of the owners participating in economically significant racing and endurance events, especially in the states of Michigan and Wisconsin. Sledding and tobogganing are enjoyed by millions, and have been so perfected by some as to warrant inclusion as an Olympic event. Ice skating and hockey are avidly pursued by many, and professional hockey is now one of the major spectator sports in North America. Additional snow and ice recreational activities include ice-boating, curling, snow-shoeing, and sled-dog racing, all of which are both socially and economically important in many areas.

The snow-based recreation industry of western Europe is even more important than that of North America, both absolutely and in relative economic terms. France alone, for example, has between 1 and 1½ million skiers (which

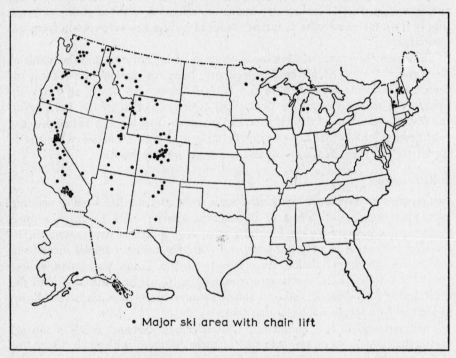

• Major ski area with chair lift

Fig. 8.III.1 National Forest ski areas in the United States (Source: U.S. Department of Agriculture, Forest Service, PA 525).

figure is expected to surpass 2 million by 1970), some 200 winter-sports resorts with 2,000 hotels containing 45,000 bedrooms. There are 40 aerial ropeways, 150 cabin or chair lifts, 800 ski lifts, 150 ski schools, and 2,000 ski instructors in the country. New centres are being opened every year, one of the most recent being the elaborate resort of La Plage in Savoy, which was begun from scratch in 1961 and can now accommodate more than 6,000 vacationers at a time. A large proportion of those taking skiing holidays in France are foreigners, with 40% of the holidays being taken at Christmas, 25% at Easter, and 35% in between the two. The economic significance of this invisible tourist export is even more important in Switzerland, where the number of nights passed by foreign tourists increased from the already-high figure of 15 million in 1912 to 19 million in 1966, the latter representing only 59% of the total tourist accommodation. In 1966 foreign tourists spent 2,900 million Swiss francs on Swiss holidays, which compares with the 8,110 million francs earned by the country's manufactured exports in 1960. One in ten of the Swiss working population is employed in transportation, hotels, and catering.

However, the aesthetic benefits of snow are difficult to assess in financial terms. How much is a White Christmas worth, or a drab winter landscape revitalized by a blanket of white? In those areas where prolonged snow-cover is common winter seems to take on a special meaning, with the pattern of life being distinct from other parts of the world. Some of these intangible assets of snow were captured in this passage from the *Atlas Maritimus* in 1728, referring to winter in Finland:

'. . . the inhabitants look abroad . . . to travel and carry on their needful affairs, and without troubling themselves about night or day, sea or land, rivers or lakes, dry land or wet, the face of the world being all smooth and white, they ride on their sledges . . . carrying a compass with them for the way, wrapt in warm furs for the weather and a bottle of aquavita for their inside, with needful store of dried bread and dried fish for their food'.

Traces of this attitude remain and help to explain the mixed reactions which still occur in response to snow. The negative impact of snow in today's urban world can be fully understood only after a careful analysis of such positive psychological associations, for it is often the aesthetic 'benefits' of snow which prevent optimum effort being directed toward its control.

3. The negative aspects of snow and ice

Perhaps the most obvious detrimental economic effect of snowfall lies in the disruption of communications and allied services, and it is at once clear that, although where snow is common authorities may spend more money in the long run in combating it, it is in regions where snowfall has a high variability that the greatest and most expensive individual disruptions occur. A classic example of snow disruption took place in Britain during the abnormally snowy winter of 1962–3. On average, southern Britain can expect not greatly in excess of 10 days with snow lying per year, the most likely period for it (2 chances in 7) being about the second week in January. Snow began to fall on 26 December and continued intermittently until 23 February, during which there were 35 consecutive days with maximum temperatures not exceeding 2·7° C, giving the coldest winter for the country as a whole since 1829–30 and for some regions since 1740. The prolonged snowfall at one time blocked 95,000 miles of road in southern England, directly caused the deaths of 49 people, and was responsible for incalculable economic loss.

Railroads provide some of the best comparative data on snow disruption, partly because their traffic is so well documented and partly because they are so susceptible to it. Besides blocking lines, freezing snow jams points and moving parts of signals, obscures signal lights, breaks telegraph wires, and falling snow may completely obscure visibility. In regions used to coping with the effects of snow costs are quite modest and the disruption at a minimum. The New York Central R.R., for example, normally spends 0·74% of its operating expenses on snow and ice removal (this figure rose to only 1·18% in 1945 as a result of the famous Buffalo–Albany blizzard), and of the normal winter train delays on the Chicago, Milwaukee, St Paul, and Pacific R.R., only 12% are due to weather conditions (mostly to blizzards, snow drifting, and snow-plough delays in the northern Rockies section). Compared with these figures, the heavy snowfalls in Britain during 1940 put 1,500 miles of track out of service for 12–72 hours, and

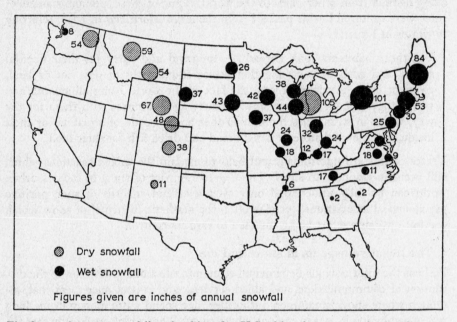

Fig. 8.III.2 Average snowfall at the thirty-five U.S. cities chosen as study sites for snow hazard. Figures give mean annual snowfall in inches (Dry snowfall<0·1 in. of water per inch of snow; wet snowfall>0·1 in. per inch).

during the bad winter of 1947 one trans-Pennine route was closed by snow for two months. Where severe snow and ice occur only occasionally, the disruption may be on a huge scale, as with the ice storm of 30 January to 2 February 1951 in Alabama, Louisana, and Arkansas, in which the principal damage was the breaking of telegraph wires (one stretch of line, for example, suffered 3,500 breaks in 111 miles) and which was not permanently restored for over two years.

An accurate assessment of the troublesome effects of snow in urban areas urgently requires some type of categorization. To this end I have separated urban snow disruption into two parts: internal, when inter-change within the city is hampered; and, external, when conditions affect the interaction between a community and its surrounding area. Measurement was based on a five-order hierarchy of disruption which categorized impact from paralysis (first order) to minimal (fifth order), with transportation the most critical variable (Table 8.III.1). Paralysis disruptions may occur in either or both of these situations. With respect to internal activity, the complete restriction of mobility with its myriad ramifications is normally the most serious problem that can be attributed to snow, since most functions characteristic of urban areas require movement from one section of the city to another. The isolation of a city from its surrounding area, or vice versa, would represent a paralysis external disruption. This type of measurement has provided considerable understanding of the disruption producing roles of the physical paramenters (the physical snow environment), and insight concerning the implications of snow-hazard perception and adjustment.

Generalizations regarding the snow hazard were based on observations, taken at thirty-five U.S. cities with annual snowfall ranging from 1 to 105 in. (fig. 8.III.2). Information concerning disruptions was primarily obtained from daily newspaper coverage and public records. From this material it was possible to classify the impact of all snow days under study during a ten-year period 1956–65.

The salient relationships between the physical snow environment and disruption can be summarized as follows:

A. Annual snowfall

  1. The frequency of higher-order (paralysis, crippling, and inconvenience) disruption increases with annual accumulation, however, not without exception. When paralysis disruptions are considered separately the correlation with annual snowfall is weak.

  2. Disruption, though it increases with annual snowfall, does so at a diminishing rate. Inversely, the intensity of disruption (disruption per inch) decreases as average annual snowfall increases.

  3. Snow of lower moisture content tends to produce less difficulty than the wet variety. Disruption per unit of snow at sites where the moisture content was lower than 0·10 per inch (chiefly cities in the Mountain West) was significantly less.

B. Individual storms

  1. At the individual storm level the relationship between depth of snow and curtailment of human activity is very strong. Furthermore, intersite comparisons revealed in general that as annual accumulation increases the amount of snow necessary to produce paralysis conditions increases as well. Thus, to take the extremes, the average paralysis disruption at Muskegon (105 in,) was associated with a 14-in. storm; whereas in Greensboro (11 in.), $4\frac{1}{2}$ in. produced those circumstances.

  2. A significant statistical difference exists between the impact of snow in association with winds of 15 miles per hour or more, and that accompanied by wind of lesser velocity.

  3. Both the time and the rate of snowfall proved critical at the single storm level. The time of fall often means the difference between a nuisance and a

TABLE 8.III.1 Hierarchy of disruptions: internal and external criteria

Activity	1st order (paralysing)	2nd order (crippling)	3rd order (inconvenience)	4th order (nuisance)	5th order (minimal)
INTERNAL					
Transportation	Few vehicles moving on city streets	Accidents at least 200% above average	Accidents at least 100% above average	Any mention	No press coverage
	City agencies on emergency alert, Police and Fire Departments available for transportation of emergency cases	Decline in number of vehicles in CBD	Traffic movement slowed	Traffic movement slowed	
		Stalled vehicles			
Retail trade	Extensive closure of retail establishments	Major drop in number of shoppers in CBD	Minor impact		No press coverage
		Mention of decreased sales			
Postponements	Civic events, cultural and athletic	Major and minor events	Minor events	Occasional	No press coverage
		Outdoor activities forced inside			
Manufacturing	Factory shutdowns Major cutbacks in production	Moderate worker absenteeism	Any absenteeism attributable to snowfall		No press coverage
Construction	Major impact on indoor and outdoor operations	Major impact on outdoor activity Moderate indoor cutbacks	Minor effect on outdoor activity	Any mention	No press coverage

Communication	Wire breakage	Overloads	Overloads	Any mention	No press coverage
Power facilities	Widespread failure	Moderate difficulties	Minor difficulties	Any mention	No press coverage
Schools	Official closure of schools / Closure of rural schools	Closure of rural schools / Major attendance drops in city schools	Attendance drops in city schools		No press coverage
EXTERNAL [a]					
Highway	Roads officially closed / Vehicles stalled	Extreme-driving-condition warning from Highway Patrol / Accidents attributed to snow and ice conditions	Hazardous-driving-condition warning from Highway Patrol / Accidents attributed to snow and ice conditions	Any mention, for example, 'slippery in spots' warning	No press coverage
Rail	Cancellation or postponement of runs for 12 hours or more / Stalled trains	Trains running 4 hours or more behind schedule	Trains behind schedule but less than 4 hours	Any mention	No press coverage
Air	Airport closure	Commercial cancellations	Light plane cancellations / Aircraft behind schedule owing to snow and ice conditions	Any mention	No press coverage

[a] Warnings are the key to this classification. They provide excellent indicators because they are widely publicized.

crippling disruption, while the rate of accumulation has a profound impact on the effectiveness of snow-control operations.

4. The effect of air temperature proved difficult to isolate, perhaps because temperature is so casually related to the amount of snow and the rate at which it accumulates.

Disruption resulting from snow and ice, like that resulting from floods or hurricanes, is only partially attributable to the physical properties of the hazard. Of equal or greater importance may be the adjustments which communities and individuals make to modify the hazard, or to reduce its impact.

TABLE 8.III.2 Results of correlation

	Coefficients of correlation					
	Magnitude of disruption (internal and external)			Intensity of disruption (internal and external)		
	1st order, paralysing	2nd order, crippling	3rd order, inconvenience	1st order, paralysing	2nd order, crippling	3rd order, inconvenience
Annual snowfall	0·23	0·73	0·87	−0·48	−0·56	−0·71
Annual snowdays	0·15	0·64	0·81	−0·48	−0·62	−0·69

1. 0·45 Significant at the 0·01 level of probability.
2. 0·35 Significant at the 0·05 level of probability.

A comparison of the disruption patterns with the snow environments of all thirty-five cities revealed some interesting relationships. A very strong positive correlation was obtained by comparing the sum of the three highest orders of disruption magnitude with annual snowfall (Table 8.III.2). Using this combination, differences in annual snowfall account for nearly 76% (i.e. $0·87^2 \times 100$) of the variation in disruption. However, by combining only the highest two orders of disruption and comparing them with snowfall, it appears that the snow environment explains only 53% (i.e. $0·73^2 \times 100$) of the variation.

The relationship between first-order disruption magnitudes only and snowfall is in marked contrast with the coefficients obtained through the summations referred to above. Paralysis situations tend to be ubiquitous regardless of mean annual snowfall or snowdays. Disruption at cities which average less than 20 in. a year is in many cases more frequent and more severe than the disruption at places recording considerably greater accumulations. For example, St Louis, Richmond, Knoxville, Louisville, Evansville, and Greensboro experience more paralysis occurrences than communities with twice to three times as much snow. To cite a specific case, St Louis (18 in.) has recorded over 60% more first-order and 40% more second-order disruptions than Milwaukee (44 in.). As a result, the relationship between paralysis and annual snowfall is statistically insignificant, accounting for less than 6% of the variation. Measurements of disruption intensity (i.e. disruption per 10 in. of snow) also provide substantiating evidence concerning these relationships (Table 8.III.2.).

A generalized pattern of snow-caused disruption in the United States based on the data from the thirty-five study sites tends to emphasize the significance

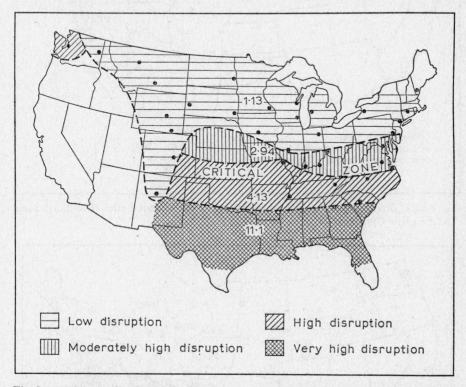

Fig. 8.III.3 A generalized pattern of snow-caused disruption in the United States. Figures indicate average number of first- and second-order disruptions per 10 in. of snowfall for the sites within each zone. Shading gives the relation of disruption to snowfall, the low disruption indicating a high degree of adjustment to it, and vice versa.

of the human environment (fig. 8.III.3). Four disruption zones are conspicuous. The northernmost zone is one marked by reliable annual snowfall, a high level of adjustment, and low disruption related to the amount of annual snowfall. It is in this area that the most sophisticated public snow-control programmes exist and where individuals have taken the initiative to protect themselves and their property from the snow hazard. It is also here where behavioural patterns have been conditioned through more frequent exposure to snow and ice, the result being a less emotional, and perhaps more routine response to most snow situations.

South of the high-exposure zone is a narrow transitional area where disruption

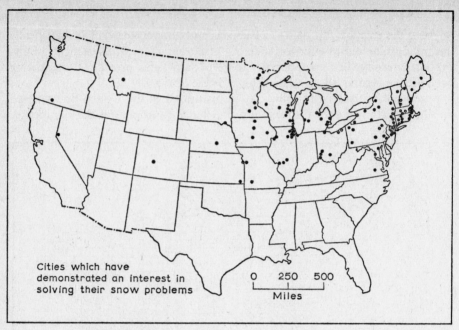

Fig. 8.III.4 Distribution of snow hazard interest in the United States (Compiled from *The American City*, 1950–67).

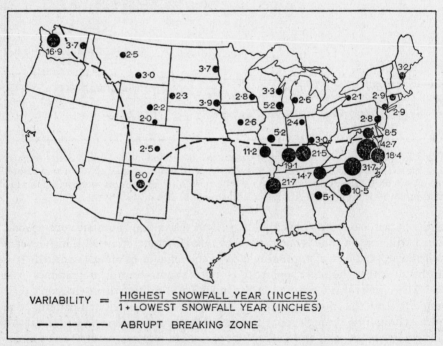

Fig. 8.III.5 Pattern of snowfall variability in the United States. Proportional circles represent the variability index and illustrate the abrupt breaking zone.

is moderately high in relation to snowfall and where inter-site variation is substantial. This zone and the wider belt to the south are characterized by marginal adjustment and relatively high disruption per inch of snow. Both areas exhibit a critical and abrupt scaling down in the level of adjustment, which has produced a general state of unpreparedness for snow. Many of the cities within this area experience a greater *absolute* number of paralysis and crippling disruptions than do their northern 'snow belt' counterparts. The fact that adjustment wanes rapidly within the transition zone is partly documented by the pattern of snow-hazard interest (fig. 8.III.4). Interest in adjustment decreased abruptly south of the east–west line approximating the 20-in. snow isohyet – the same area in which both the disruption per unit of snow and the annual snow variability increase!

Annual variability of snow may well explain much of the observed patterns of adjustment and disruption (fig. 8.III.5). A comparison of expected variability over a ten-year period for cities with accumulations between 10 and 20 in. points out the degree of uncertainty associated with low-snowfall areas. Cities in this 'normally' meagre to moderate snowfall zone are confronted with a degree of uncertainty unknown in the north. Here the decision-makers must cope with a snow environment characterized by a 2- or 3-in. accumulation in one year and as much as 30–40 in. in the next. It is clear that community adjustment is not a direct function of annual snowfall, but rather there seems to be a critical accumulation to which cities respond that is explained largely by community decision-making and perception.

In conclusion, the pattern of snow-caused disruption in the United States suggests the immediate need to systematize decision-making in regard to snow-control programmes. It appears that planning has generally been formulated in a haphazard manner, with appropriations being allocated via the political marketplace. Even in high snow-exposure areas planning has been geared to some preconceived normal environment. As a result, high-snowfall cities are not equipped to cope with the five-, ten-, and fifty-year storms, while low-exposure communities have often embraced 'hope and pray' strategies. A substantial research effort is needed to identify the costs of snow-related disruption before the benefits of various snow-control alternatives can be assessed.

REFERENCES

AMBASSADE DE FRANCE [1968], *The French Tourist Industry* (Service Presse et d'Information, London).

ANDERSON, H. W. [1963], Managing California snow lands for water; *U.S. Forest Service Research Paper* (Berkeley, California).

ANDERSON, H. W. [1966], Integrating snow zone management with basin management; In Kneese, A. V. and Smith, S. C., Editors, *Water Research* (Johns Hopkins University Press, Baltimore), pp. 355–73.

BAUER, H. [1968], *All About Switzerland* (Swiss National Tourist Office).

BELL, C. [1957], *The Wonder of Snow* (New York), 269 p.

CHAMPION, D. I. [1947], Weather and railway operation in Britain; *Weather*, **2**, 373–80.

HAY, W. W. [1957], Effects of weather on railroad operation, maintenance, and construction; *Meteorological Monographs*, **2** (9), 10–36.

LIEBERS, A. [1963], *The Complete Book of Winter Sports* (New York), 228 p.

MEAD, W. R. and SMEDS, H. [1967], *Winter in Finland* (Hugh Evelyn, London), 144 p.

ROONEY, J. F. JR. [1965], *The Urban Snow Hazard: An Analysis of the Disruptive Impact of Snowfall in Ten Western and Central United States Cities* (Clark University, Worcester, Massachusetts), 150 p. (Also from University Microfilms, Ann Arbor, Michigan.)

ROONEY, J. F. JR. [1967], The urban snow hazard in the United States: An appraisal of disruption; *Geographical Review*, **57**, 538–59.

RUSSELL, J. A., Editor [1957], Industrial operations under extremes of weather; *Meteorological Monographs* **2** (9).

THOMPSON, J. C. [1959], The snow probability factor; *The American City*, **74**, 80–3.

9.III. Human Response to Floods

W. R. DERRICK SEWELL

Department of Economics, University of Victoria

1. Man's affinity for floodplain occupance

One of the most conspicuous features of human settlement pattern is man's affinity for riverine locations. Throughout history he has been attracted to the lands adjacent to rivers. Today a very considerable proportion of the world's population lives in such areas. There are some fairly obvious reasons why this should be so. River valleys often contain deposits of rich alluvium, providing the basis for the development of a thriving agricultural industry. Some of the world's great civilizations have developed in the bottom lands of major rivers, notably along the banks of the Tigris and Euphrates, the Nile, the Indus, and the Yangtze. River valleys are often transportation corridors, providing access for roads and railways. They also provide level land for the construction of houses and factories. For certain activities a riverine location is essential, particularly those which depend upon the river for transportation or for large quantities of water for processing. Other activities which have a greater freedom of locational choice may also be attracted to the river because of the activities already there. Various aesthetic considerations also provide an attractive force. Riverine locations carry a prestige value, for example, for private homes.

Settlement beside a river, however, can be a mixed blessing, for once in a while the river may overflow its banks and exact a heavy toll of property losses, income losses, and sometimes losses of life as well. In some cases man has learned to live with such periodic inundations of the floodplain and has turned them to economic advantage. The case of the Nile is perhaps the most famous. 'Egypt,' said Herodotus some 2,400 years ago, 'is the gift of the Nile.' His statement is still true today. In late June each year the lower Nile, swollen by tropical rains and the melting snows of its upper reaches, begins to rise. By late September its floodplain has become a large lake. As the floodwaters begin to recede in October they deposit a rich residue of silt, which revitalizes the soil. Through the construction of a network of canals, dykes, reservoirs, and ditches, the Egyptians have developed an agricultural system which is geared to the annual inundation of the Nile's floodplain.

In the case of the Nile floods are a critical input into the economy. In most cases, however, floods are regarded as a burden rather than as an advantage. Often they cause huge losses of property and income, and sometimes large losses of life as well. There are records of floods in China which have caused more than 1 million deaths at a time. Property losses can also be staggering. In 1951, for example, the Kansas River in the United States overflowed its banks and caused damages exceeding $1·5 billion. Flood losses in the United States have exceeded $1 billion several times in recent years. Effects of floods in the lesser-developed countries are even more serious, for recovery is much more difficult. Several times in recent years, for example, hundreds of thousands of people have been left homeless in India, Pakistan, Korea, and China, and their sources of food and livelihood have been severely damaged.

Despite the huge losses that have been experienced, floods have not discouraged settlement in river valleys. On the contrary, there is substantial evidence that occupance of floodplains in many parts of the world is increasing. The Yellow River and the Yangtze River in China have overflowed their banks many times in the past four thousand years, and millions of people have been drowned as a result, yet the peasants continue to flock into the floodplain. In this case the reasons are fairly obvious: there are few alternative opportunities for earning a living or for growing the food needed by a burgeoning population.

Occupance seems to be increasing, however, even in those cases where economic necessity is not involved. In some instances the costs of living or working in a place outside the floodplain might be lower than for those living on the floodplain itself, yet others flock in to join them. In the United States it has been estimated that at least 12% of the population lives in areas subject to periodic inundation (White *et al.*, 1958). A similar proportion of the Canadian population also lives in floodplains (Sewell, 1965). In both cases the proportion seems to be increasing. Floodplain occupance seems to be growing at a more rapid rate than overall population increase. Flood losses, therefore, seem destined to continue to mount.

Given these trends, what are the possible courses of action that can be taken to reduce the impending toll of flood losses?

2. Possible adjustments to floods

There are several possible adjustments to floods. Briefly, these might be grouped into the following categories:

A. Accepting the loss
B. Public relief
C. Emergency action
D. Structural changes
E. Flood proofing
F. Regulation of land use
G. Flood insurance
H. Flood control.

Each of these adjustments has advantages and disadvantages. Each of them is appropriate for some situations but not for all. Generally several adjustments are tried before a final selection is made. Often a combination of several adjustments is chosen. Some of the adjustments can be adopted by individuals: others, however, depend upon group action. What are the characteristics of the various adjustments, in what circumstances are they most appropriate, and what are the implications of their adoption for public policy?

A. Accepting the loss

The most common adjustment, perhaps, is to accept the loss. This is certainly the case where people are too poor to do anything else or are unaware of any alternative course of action. In the more developed countries, however, it is seldom the result of a conscious decision, particularly when a flood has been experienced in the past. Usually an attempt will be made to find means for offsetting future losses.

B. Public relief

One of the more common alternatives to accepting the loss is to rely upon public relief. An immediate reaction to the announcement of a flood disaster is the establishment of a relief fund to assist flood victims. Sometimes such funds are purely voluntary, and calls are made for contributions from people in adjacent communities, from the country at large, or even overseas. Often these voluntary relief funds gather very substantial amounts of money. Another type of relief fund is provided by the Government. Typically there is no set policy for determining the amount to be granted, and often there is considerable debate on the matter. A third type of relief is that administered by the Red Cross and other similar voluntary organizations. Friends and relatives may also offer assistance to flood victims, providing them with food and temporary accommodation.

The principal justification for the various forms of relief is that they help to ease the immediate distress, and to aid the initial rehabilitation. It is sometimes a very useful adjustment to floods. Generally, however, it tends to become regarded as a right rather than as a charitable gift. As a consequence, it tends to remove the incentive to avoid future flood losses, and therefore encourages persistent human occupance of the floodplain.

In the United States the Federal Government has allocated funds for relief of flood victims, through gifts, low-cost loans, deferred payments, and subsidies of various kinds. Similar types of flood relief have been provided in Canada and in several other countries. Results of studies in North America seem to suggest that the granting of public relief in the past constitutes a major reason for persistent human occupance of floodplains, particularly when the provision of such relief is made without any obligation on the part of the recipient to undertake measures to reduce his vulnerability to future flood losses (Sewell [1965], p. 70).

C. Emergency action and rescheduling

Potential losses of property and income can be reduced by various types of emergency action or by rescheduling of activities. Emergency action consists mainly of removing persons or property from the area subject to flooding, and flood fighting. Some communities have come to rely almost exclusively on this

form of action. Each year when the floodwaters begin to rise they make prepara-
tions to evacuate the area. Sometimes the actual evacuation is undertaken by the
individuals themselves, but generally the local authority or the central government assumes responsibility for overall organization. Massive evacuation programmes are undertaken along the Lower Mekong, the Indus, the Mississippi,
and other major rivers each year.

Efforts are also made to reduce the impact of floods by flood fighting, such as
by building temporary dykes along the river or outside a building, elevating
goods and equipment from the reach of the floodwaters (such as by removing
them to an upper floor of the building), or by protecting equipment with plastic
sheeting or grease. Here again there are opportunities for both individual action
and group action. Generally, however, action taken beyond the individual's
dwelling or work-place depends upon organization by government authorities.

Another form of emergency action that is undertaken in some places is the
rescheduling of operations. Business managers, for example, may so organize
their production schedules that they avoid having damageable goods in the flood-
prone area at the time when the flood is expected. In some cases they may
schedule vacations to avoid losses of income caused by the closing of the plant.
Rescheduling can be carried out fairly easily in some activities, but not in all.
Many types of agricultural activity, for example, are restricted as to the time
they can be undertaken, although there is sometimes a certain amount of latitude
as to the timing of planting or harvesting. There are often opportunities in
transportation and in manufacturing industries, however, for rescheduling.
Potential losses can be reduced, for example, by making arrangements for send-
ing passengers and goods over routes outside the floodplain, and by arranging for
production of certain goods and services at plants outside the floodplain during
the flood period.

The effectiveness of emergency action usually depends upon the extent of
preparation before the flood occurs. The floodplain occupant must know how
probable it is that there will be a flood and what its effects are likely to be. They
must also be aware of the kinds of emergency action they might take to reduce
potential losses. In addition, they must be given adequate warning of the onset
of the floods to enable them to put their plans into operation. Key elements in
emergency action, therefore, are the provision of information about the likelihood
and the potential effects of floods, and the development of a flood-forecasting and
warning system.

Emergency action seems to be an appropriate adjustment in those situations
where the flood-to-peak interval is greater than one hour. (The flood-to-peak
interval is the period from the time the river reaches flood stage to the time it
reaches its maximum stage.) When this interval is less than 1 hour only limited
kinds of emergency action can be taken. When it is greater than 1 hour many
opportunities for reducing losses present themselves. Very substantial opportunities exist when it is more than one day. In the case of the 1952 flood on the
upper Mississippi, for example, the flood-to-peak interval was 180 hours ($7\frac{1}{2}$
days). Many floodplain dwellers took action to reduce potential losses, such as

disconnecting utility lines, raising their houses above flood level by jacking them up and supporting them on concrete-block piers, removing belongings to upper floors, and sealing windows and doors with temporary barricades.

Emergency action can substantially reduce potential flood losses. In the United States, for example, it has been estimated that such action taken following flood warnings appear to have reduced losses by at least 5% and sometimes by as much as 15% (White [1939]; U.S. Select Committee, [1959], p. 7). Such action, however, is not a panacea. It is most effective when the flood duration is short, where flood velocity is low, and where the frequency of flooding is high. Interest in emergency action tends to lag when the interval between floods is long. Its success hinges upon floodplain occupants being able to interpret information about floods and their being able to select the appropriate adjustment. There is evidence that floodplain occupants are not always able to do so efficiently. Moreover, there is sometimes resistance to taking any action, even though a flood warning has been issued.

Like public relief, emergency action tends to encourage persistent human occupance of floodplains. It has the advantage, however, that it encourages individuals to take action to reduce personal losses.

D. Structural changes

Another way of reducing potential flood losses is to modify building structures to repel floodwaters. Among the various types of structural change are construction of walls with impervious materials, closure of low-level windows and other openings, and underpinning of buildings. In some cases buildings can be built on stilts. This enables buildings to perform several functions. Cars can be stored between the stilts and moved during the flood period. In certain instances land fill is a practical proposition. On Annacis Island in British Columbia a massive programme of land fill has been undertaken to reduce potential losses from flooding of the Fraser River. Factories have been built upon this fill.

Structural change is appropriate where the duration of flooding is short and where the velocity is low. It appears to be most effective when the depth of flooding is less than 3 ft. It is possible, however, to build structures to withstand depths of floodwater in excess of 15 ft. Modifications can be made to existing structures, or they can be incorporated into new ones. It is usually much less expensive, however, to build them into new structures. Land fill can help to reduce the impact of flooding. It is an especially attractive adjustment when undertaken prior to construction in an urban or industrial area. It generally becomes prohibitively expensive once such development has taken place.

Structural change and land elevation can be undertaken by individuals or by groups. Thus far governments have played only a minor role in encouraging the adoption of these adjustments. There are signs, however, that they are likely to become fairly widespread in the next few years in the United States, particularly as a result of changes in Federal Government policies relating to flood management (U.S. Government, 1966). Various incentives might be offered to encourage the adoption of these measures, including grants of low-cost loans to those who are willing to undertake structural change or the withholding of mortgages unless the structure is built to withstand a specified flood.

Structural change and land elevation tend to encourage persistent human occupance. They do offer, however, a means of reducing potential losses, and they do place part of the burden on the floodplain occupant.

E. Flood proofing

Flood proofing is essentially a combination of structural change and emergency action. It does not necessarily involve evacuation. Rather it concentrates on the adoption of certain measures that can be put into action as soon as a flood warning is received. Among the various types of flood proofing are the installation of removable covers, such as steel or aluminium bulkheads, over doors or windows, or the installation of sump pumps and elevated outlet pipes to remove water which seeps into basements and interiors of buildings. In stores counters can be placed on wheels to facilitate rapid removal. There are, of course, many other possibilities for flood proofing (Schaeffer, 1960).

Flood proofing offers considerable opportunities for reducing flood losses. Many factories and stores in the Golden Triangle of Pittsburgh, for example, have adopted this adjustment and have found it a very effective means of dealing with floods. Flood proofing is now an integral part of the TVA flood-control programme. To be effective, however, it requires a well-organized flood information system. Like other adjustive actions to floods, flood proofing tends to foster persistent human occupance of floodplains, yet similar to emergency action and rescheduling and structural changes, it does place part of the responsibility for taking action on the shoulders of the individual.

F. Regulation of land use

The land in the floodplain has a wide variety of potential uses, ranging from urban and industrial development, through agricultural and recreational uses, to leaving it in its virgin state. Potential losses tend to vary with the type of use, being highest on land used for urban or industrial purposes, and lowest on land set aside for agriculture or recreation. There would be no losses, of course, on land that is not used at all. The immediate conclusion might be to keep all development out of the floodplain. This would be neither realistic nor necessarily the economically most sensible course of action. Ideally, an attempt should be made to determine which activities can afford to locate in the floodplain and still pay the 'natural tax' of flood losses (Renshaw, 1961). If the activity cannot afford the latter, then it should not be allowed to use the floodplain. A set of regulations based upon this concept would encourage potential floodplain occupants to examine carefully locations outside the floodplain as well as inside the floodplain.

Figure 9.III.1 illustrates the manner in which floodplain land might be allocated among competing alternative uses. Each use has a rent-earning capability, determined by the returns it can produce after paying the costs of hiring the various factors of production. Assuming that the total land in a given floodplain is *OW* acres, and there are three competing uses, how can the area be allocated efficiently among them? The lines *FQ*, *DU*, and *BW* indicate the returns that could be earned in urban, agricultural, and recreational uses, respectively. On this basis *ON* acres would be used for urban purposes, *NS* for agriculture, and *SW* for recreation.

Flood losses, however, constitute a cost and must be taken into account in calculating net returns. Such losses result in a reduction of net returns, as represented by the dotted lines, *EP*, *CT*, and *AV*. It will be observed that urban land use is reduced to *OM*, agricultural land use to *MR*, and recreational use to *RV*. *VW* acres are then abandoned. This indicates that once flood losses

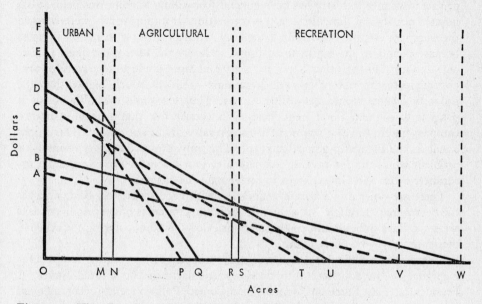

Fig. 9.III.1 The allocation of floodplain land among competing uses.

are taken into account, some urban occupants would move elsewhere rather than absorb the loss, and that land shifts into uses where losses per acre are less.

There are a wide variety of means of regulating floodplain occupance, notably through statutes, ordinances, subdivision regulations, government purchase of property, and subsidized relocation. Each of these methods have been used to varying degrees in North America and elsewhere (Murphy, 1958). Generally, government action is required to formulate and enforce the regulations. In some cases local authorities have enacted such regulations, but it often requires the action of a senior government to make this type of adjustment effective. Local authorities generally hesitate to enact regulations because they fear a neighbouring municipality will not do so, and therefore will attract activities which might otherwise have located in the first municipality. Co-operation between local authorities or central government enactment and enforcement appears to be a prerequisite for successful regulation.

Regulation of land use has a number of advantages. The most important perhaps is that it encourages careful weighing of the costs against the benefits of floodplain occupance. It forces consideration of the relative advantages of being in the floodplain versus location elsewhere. It offers a valuable complement to other types of adjustment, such as emergency action, flood proofing, structural change, flood control, and flood insurance.

G. Flood insurance

Another possible response to flood problems is to insure against the losses which they cause. Thus far, however, it has been adopted to only a minor extent. The private insurance industry has been generally unwilling to enter this field. It has pointed out several difficulties in this connection. If uniform rates were charged the company would find itself loaded up with an adverse selection of risks because people in the highly flood-prone areas are the ones most likely to take out a policy. On the other hand, if an attempt were made to charge rates proportionate to the risk of loss the premiums would be much higher than the property owners would be willing to pay. They also note that although it is possible to estimate flood frequencies, it is conceivable that a given insurance company might have to pay out claims several years in succession. This might result in the company going bankrupt. The only way of hedging against this problem would be for several companies to join forces in providing flood insurance, or for the Government to underwrite the scheme.

There are other difficulties in connection with flood insurance. Basic data for the determination of fair and equitable premiums for areas of varying flood risks are sometimes difficult and costly to obtain. Costs of administering the scheme, therefore, might be fairly high.

Despite these difficulties, however, there appears to be growing support for insurance as an adjustment to floods, particularly in the United States. A Presidential Task Force on Federal Flood Control Policy recently recommended that serious consideration be given to flood insurance, sponsored if necessary, by the Federal Government (U.S. Government, 1966). The Task Group noted the advantages of such insurance, particularly the fact that it provides an incentive for floodplain occupants to reduce damage potentials. In this way they would reduce their premiums. Flood insurance also shifts the burden of flood losses on to those who are responsible for them, the floodplain occupants.

Flood insurance tends to encourage increased occupance of the floodplain, but it does so selectively. Only those activities that can afford to pay the premiums can afford to continue occupancy (Krutilla, 1966).

H. Flood control

Finally, man may adjust to floods by trying to control them. Two main lines of action are possible: one in the land phase (flood abatement) and the other in the channel phase (flood protection) (Hoyt and Langbein, 1955). Examples of flood abatement are the modification of cropping practices, terracing, gully control, bank stabilization, and revegetation. In the United States the Department of

Agriculture has undertaken major programmes of this type aimed at controlling the development of floods. Projects focused on forest replanting, soil-erosion control, and improvement of farming methods have been carried on with flood abatement as one of their major objectives. Many millions of dollars are allocated each year by the Federal Government for this purpose. Similar programmes have been sponsored in Canada by the Prairie Farm Rehabilitation Administration, the Maritimes Marshlands Rehabilitation Administration, the Department of Agriculture, and the Department of Forestry, as well as by various provincial government agencies. Generally, flood control is one of several objectives of these programmes. Typically, other objectives include agricultural readjustment, soil conservation, and the preservation of fauna and flora.

Flood-protection programmes are concerned with the channel phase of floods. Their objectives are to control the flood once it has formed and to minimize the damage it causes by regulating its flow or directing it away from damageable property. It may involve the construction of control works, such as dykes, floodwalls, or dams and reservoirs, or the undertaking of channel improvements and dredging. Flood protection is one of the most widespread of all adjustments to floods, both in the more advanced and in the lesser developed countries.

In some cases the provision of flood protection is allied with the development of projects for other purposes. Flood control, for example, is one of several objectives of the Tennessee Valley Authority Scheme, as it is in the Mississippi River and the Columbia River schemes. In this way flood control can often be provided at a much lower cost than if it were furnished on a single-purpose basis.

For some years there has been a controversy in the United States as to whether flood abatement or flood protection is the more efficient adjustment to floods. This has come to be referred to as the 'upstream–downstream' controversy, or the conflict between 'big dams and little dams' (Leopold and Maddock, 1954). On the one side are those who argue that the best way to deal with floods is to control them where they form. They suggest that deforestation and devegetation are major causes of floods, and so the most appropriate course of action is to plant trees in the place of those that have been cut down, to improve the vegetative cover in areas subject to soil erosion, and to phase out farming practices that contribute to such erosion. They see dams as a possible means of controlling runoff but feel that these are most effective in the headwaters rather than downstream. They point out further in this connection that upstream reservoirs are less likely to take out of production agricultural land or forest land. The most vigorous supporters of these views have been the Department of Agriculture, particularly through its Forest Service and its Soil Conservation Service, and various private conservation groups.

On the other side are those who believe that the most efficient way to deal with floods is to control them close to the place where they are likely to do the most damage. They point out that it takes several reservoirs in the upstream region to do the work that one large reservoir farther downstream might do. They also note that the contributions of programmes of reforestation and soil conservation to runoff control may be insufficient to deal with the flood problem. The major benefits of such programmes, they claim, are in the regions where they are undertaken. Big dams and various downstream control works are seen as the much more effective alternatives. The major proponents of the latter view are the U.S. Corps of Engineers, an agency with a long tradition of constructing large-scale engineering works to deal with flood problems.

There is some truth in both sets of arguments. Flood abatement and flood protection are not necessarily alternatives. Often they can be complementary parts of an overall programme of adjustment to floods. It is worth noting in this connection that the Department of Agriculture in the United States has recently increased its emphasis on engineering works in its programmes, and it is now building big dams as well as small ones. By the same token, the Corps of Engineers, is tending to encourage the adoption of adjustments in addition to engineering works in dealing with flood problems.

Flood abatement and flood protection can be undertaken by individuals, but generally they require co-operative action, and usually government sponsorship. Both tend to foster increased human occupance of the floodplain. Flood protection in particular tends to develop a false sense of security among floodplain occupants. Floodplain occupants may take the construction of a dyke or dam to mean that there will never be any more flooding. Consequently, more and more people move in, and activity in the floodplain intensifies. When a flood of greater magnitude than that which the dykes or reservoirs were designed to control eventually comes along much greater damage is done than if no protection had been provided at all. Flood protection, therefore, needs to be supplemented by other measures which control the increase of potential flood losses.

3. Limitations to the range of choice

The types of response to floods outlined in the foregoing discussion constitute a Theoretical Range of Choice from which a flood manager could select an appropriate course of action (White, 1964). Typically, however, only a few of these possibilities are taken into account in flood-management decision-making. The result may be that adjustment is much less efficient than it could be if the whole range was considered.

One of the major foci of research on flood problems in recent years, particularly at the University of Chicago under the leadership of Gilbert White, has been the factors which tend to limit the range of choice. Results of these studies suggest that two main sets of factors are involved: the flood manager's perception of the nature and magnitude of the flood problem and his perception of alternative responses to that problem; and various social guides which tend to encourage the consideration of some responses and to discourage the consideration of others. Kates [1962] has shown that there are wide differences in the perceptions of individual floodplain occupants as to the nature and magnitude of the flood problem, and that their perceptions often differ considerably from those of the engineer or technician. He notes that floodplain occupants often perceive the flood hazard and its potential effects rather imperfectly. As one might expect, those who have experienced a flood in the area in the past tend to have more accurate perceptions of the hazard than those who have not had such experience. But it does not necessarily follow that even when there is accurate perception of the hazard that there will be effective action to deal with it. Some floodplain occupants may feel that they will not suffer any damage in the future, and even if they do, it will not be serious. From the evidence gathered so far it seems that action is most likely when several flood events have been experienced and when the losses have been severe. For the most part there tends to be apathy about the flood hazard. Action seems to await a crisis to provide the necessary trigger.

One reason why the floodplain occupant is often unconcerned about flood problems is that they may be only one of many problems that concern him in his daily life. As a result, he devotes only a small part of his time to dealing with such matters. Typically, he is unaware of the wide range of actions he can take to reduce potential flood losses. Often he places great faith in adjustments that are considered to be of limited value by the technician, such as the clearing of brush or debris from the river channel or last-minute flood-fighting efforts.

Various social guides, such as law, historical precedent, jurisdictional constraints, and government policies, condition to an important extent the adjustments that are chosen. Typically, the questions will be asked – what has been done in the past, and whose responsibility should it be – rather than what is the best course of action in this case? Sometimes legislation facilitating government action is drafted in fairly restrictive terms, focusing upon only one or two types of adjustment. In some cases only flood relief or flood protection are seen as possible responses to flood situations, and these are written into the legislation. Consequently, such other alternatives as flood insurance, flood proofing, etc., may not be considered at all in actual decision-making, even though they may be much more effective in dealing with flood problems than the latter.

Gilbert White and his colleagues have concluded from their studies that such limitations on the range of choice account in part for the continuous increase in flood losses in the United States. More than $7 billion has been spent on various adjustments in the United States since 1936, but flood losses have continued to mount. Government policy and historical precedent have tended to favour a concentration on flood protection, emergency action, and flood-relief payments. Unfortunately, these adjustments attack the effects rather than the causes of flood losses. Generally too, they have tended to encourage a transfer of responsibility for action from the individual to the public at large.

A tangible result of the studies has been some significant shifts in U.S. government policy relating to flood problems. Important among these shifts have been a broadening of the approach to include adjustments in addition to flood prevention, emergency action, and flood relief. Consideration is now being given to policies which would broaden the practical range of choice to include flood proofing, structural change, flood insurance, and other alternatives that have promise of reducing flood losses. Particular attention is also being paid to courses of action which would encourage private individuals to consider the risks of floodplain occupance and to take appropriate measures to curb flood losses.

4. The Lower Fraser Valley: a case study

The Lower Fraser Valley in British Columbia provides an opportunity to examine the changing pattern of responses to a major flood problem. For more than a century this 800-square-mile, wedge-shaped area has been the main focus of settlement and economic development in the province. Today it provides a home for more than one million people, and it contains the larger part of the province's manufacturing, commercial, and agricultural activity. It is also a

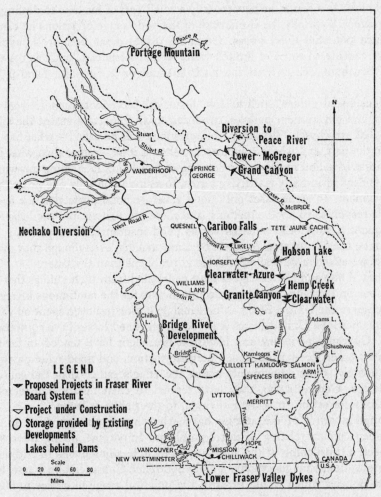

Fig. 9.III.2 The Fraser River basin. Scheme of hydroelectric power and flood control recommended by the Fraser River Board.

major transportation corridor, linking the Pacific Coast with eastern Canada, and northern North America with the United States (fig. 9.III.2).

A considerable portion of this settlement and economic activity is located in the floodplain of the Valley. More than 10% of the Valley's population, most of its agricultural activity, a large segment of its manufacturing and commercial activity, two transcontinental railways, and the major highways are in areas that are subject to inundation by major floods.

The concentration of population and economic development in the floodplain is in part a reflection of the economic advantages it offers, particularly its fertile alluvial soil, its flat terrain, and its access to other areas. It does not follow, however, that the present pattern is a result of a conscious weighing of these advantages against the costs of floodplain location. On the contrary, it is probable that many of the people who live there are unaware of the flood hazard, and so do not take it into account in their decisions. Occasionally, however, there is a major flood and the costs of floodplain occupance are brought home to them in dramatic, and sometimes catastrophic, fashion. This was indeed the case in 1948.

A. The 1948 Flood

The 1948 Flood was by far the worst disaster in British Columbia's history (Hutchinson, 1950). It was also one of the most costly floods ever to occur in Canada. There had been a heavy snow pack during the previous winter, and the spring was late in coming. Suddenly at the beginning of May temperatures in the Interior of the province began to rise, and they remained high for several weeks. The snow began to melt rapidly, and this resulted in disastrous flood conditions downstream. The danger mark on the Mission gauge, some 50 miles from the mouth of the river is 20 ft. By 25 May the river had reached 18·8 ft on the gauge (Fraser River Board, 1963). The next day the dykes at Mission broke. There were breaks at several other points downstream too. For more than thirty days the Valley was in a state of siege by the flood waters. Vancouver was cut off from the rest of Canada, except by air.

The flood resulted in huge losses of property and income throughout the Valley. No complete assessment of these losses was made, but those for which compensation was paid totalled more than $20 million. Some 55,000 acres of agricultural land were inundated. Farm buildings were damaged or destroyed. Fields in the Valley were covered with several feet of mud and debris. More than 2,000 people were left homeless. The effects of the flood extended far beyond the Valley itself, particularly because of the interruption of communications. Recovery was slow, and it was costly. Residents and politicians were resolved that such a disaster must never occur again. A thorough investigation of the flood problem was called for, leading to a set of recommendations for action.

B. Evolution of adjustment to floods in the Lower Fraser Valley

The early settlers in the Lower Fraser Valley experienced several severe floods. Some of them moved elsewhere, others decided to bear the loss in the hope that they would not be inundated again. A few, however, built dyking systems (Sewell, 1965). The first dykes were built by private landowners to protect their property. Later a number of Dyking Districts were formed to provide flood protection on a communal basis. The early dykes proved to be a disappointment, technically and financially. Many of them collapsed, and most of them fell into disrepair because it became impossible to collect enough money to maintain them. Eventually the Government felt obliged to take over the dyking system, to maintain it and to extend it where necessary. But public ownership did not solve the financial difficulties nor did it overcome the technical problems. The debt continued to mount and the dykes continued to fall down. The poor condition of the dyking system was in large part responsible for the huge losses that occurred in the 1948 Flood.

Besides the construction of flood-protection works, attempts were made to deal with the flood problem by emergency action. Individuals and communities had devised methods of flood fighting and evacuation over the years, and these had often proved to be effective, particularly in minor, short-term inundations. Evacuation and flood fighting in the 1948 Flood, however, were required on such a large scale that they could not be left to private initiative. Carefully planned and co-ordinated efforts were clearly required. Responsibility for organizing evacuation and temporary relief was assumed by the provincial and federal governments.

Up to 1948 flood relief had been a matter that was left largely to private initiative. flood victims either drew on their own resources or depended upon assistance from relatives and friends. In 1948, however, the senior governments decided that such aid would be inadequate to facilitate recovery and offered assistance for relief and rehabilitation.

Adjustment to floods in the Lower Fraser Valley up to 1948, therefore, was characterized by the adoption of relatively few measures, mainly bearing the loss, flood protection, and emergency action. Over the years there was a shift in responsibility from the private individual to the Government. The relative infrequency of major floods had led to a relaxing of vigilance on the maintenance of the dyking system. Nothing was done to control occupance of the floodplains.

The 1948 Flood led to some important shifts in public policy. First, the Federal and Provincial Governments offered financial assistance on a fairly large scale for relief and rehabilitation. Although no promise was made that such aid would be forthcoming in the future, it is evident that this established a precedent. Many of those who now live in the floodplain believe that financial assistance would be provided in the event of another flood. Second, the two governments agreed to search for a more permanent solution to the flood problem. As an initial step they established in 1949 an engineering board – known initially as the Dominion-Provincial Board, Fraser River Basin, and subsequently as the Fraser River Board – to undertake a thorough investigation of the causes of the problem and to recommend measures that should be adopted to deal with it. It presented its Final Report in 1963.

C. The Fraser River Board's proposed scheme

The Fraser River Board proposed as a solution to the flood problem the construction of a scheme of hydroelectric power and flood-control development, estimated to cost about $400 million, and the reconstruction of the dyking system in the Lower Fraser Valley, costing some $5 million (fig. 9.III.3). The scheme would control a Design Flood, which has a discharge of some 600,000 ft³/sec at Hope (fig. 9.III.4) and a chance of occurrence of one in 150 years (fig. 9.III.5) This control would result in a reduction of potential flood losses, valued at some $75·3 million per annum (fig. 9.III.6). The scheme would also generate about 785,000 kW of firm power. Revenues from the sale of this power would be sufficient to cover the entire cost of the scheme. The only costs of controlling the floods, therefore, would be the costs of repairing the dyking system.

So far little has been done to implement the Board's recommendations. There have been discussions between the Federal Government and the British Columbia Government, but no agreement has yet been reached. Even if an agreement had been reached to construct the proposed scheme, however, it is not clear that it would lead to a long-run reduction in flood losses in the Lower Fraser Valley.

Three main factors have delayed the implementation of the Fraser River Board's proposed scheme: the uncertainty as to which level of government is responsible for initiating action to deal with flood problems; the pre-emption of the power market by other projects under construction in British Columbia; and opposition from recreation interests. Traditionally, flood problems have been regarded as a matter of local responsibility in British Columbia. In 1948, however, the senior governments indicated that they were willing to give assistance in dealing with such problems, both by providing funds for relief of flood victims and by undertaking an engineering investigation. But no commitment was made by either the Federal or Provincial Government beyond that. The Federal Government feels it is unable to initiate action because flood problems are beyond the jurisdiction granted to it under the British North America Act. The Provincial Government, however, feels that the proposed solution is beyond its financial capabilities. Moreover, the local authorities seem unwilling to take any remedial action themselves because they feel the senior governments have assumed overall responsibility for dealing with it!

Another reason for the delay in action has been the pre-emption of power markets. While the studies of the Fraser River were being undertaken, investigations of other rivers in the province were also under way, notably those relating to the Columbia River and the Peace River. Huge hydroelectric power schemes are now being built on the latter rivers which will provide power for the Province and for export to the United States for the next fifteen to twenty years. Unfortunately, the planning relating to the Fraser River was not phased in with that relating to the other two rivers, and so the possibilities of an integrated scheme involving all of them were not considered.

The proposed Fraser River scheme has also been delayed because of opposition

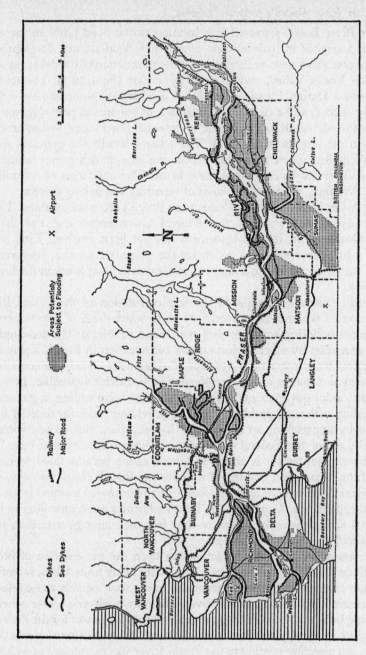

Fig. 9.III.3 The lower Fraser Valley dyking systems.

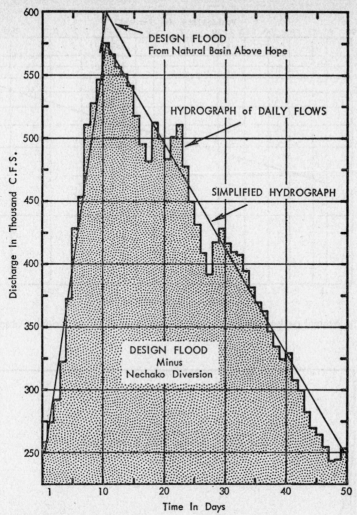

Fig. 9.III.4 Design flood hydrograph at Hope, British Columbia.

by salmon fishing and recreation interests. The river is one of the world's largest remaining sources of salmon. Salmon fishing is one of the oldest industries in the province and still makes an important contribution to employment and income in British Columbia. The landed value of the catch from the Fraser River is about $20 million per annum. Some 10,000 fishermen and shore workers are employed on a full-time or part-time basis in obtaining the catch and processing it. In addition, there is a rapidly growing sports fishery in the province which provides recreation opportunities and an increasing source of income for vendors of fishing equipment and supplies and for guides.

Construction of dams on the Fraser River would interfere with the migration and spawning of the salmon runs, and so the commercial fishermen and the

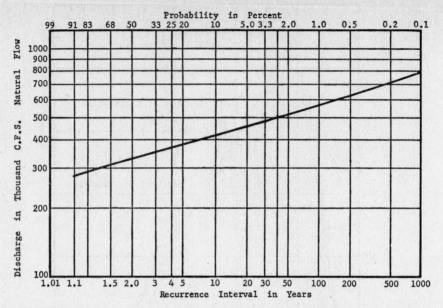

Fig. 9.III.5 Annual peak discharges of the Fraser River at Hope, British Columbia.

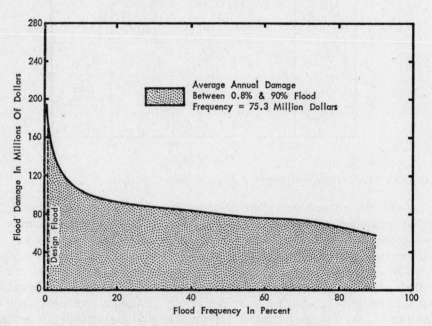

Fig. 9.III.6 Damage frequency curve for the lower Fraser Valley.

sports fishermen have been solidly against proposed adjustments to the flood problem involving such structures. In an effort to accommodate this opposition, the Fraser River Board decided to restrict its selection of sites for dam construction to streams in the headwaters of the river where there would be only minimal effects on the salmon runs. Such a location resulted in a scheme that was considerably less efficient both technically and economically than others that could have been developed. The storage reservoirs are far from the major areas of flooding, thus requiring a larger number of projects than would have been the case with near storage. The power plants associated with them are also far from major load centres, and there are no opportunities for large-scale economies of power production at the projects selected. Power costs, therefore, are not competitive with those of other sources in the province.

Despite the efforts of the Board to accommodate the fishing interests, there has been opposition to the proposed scheme. Some of them claim that the construction of any structure will have adverse effects elsewhere in the river system, and therefore should not be allowed. There is also opposition from recreation interests, who feel that the construction of some of the proposed reservoirs will destroy 'the unique natural beauty' of the areas involved. They are particularly opposed to the projects recommended for the Clearwater River. Unfortunately, these are among the most economical of those recommended by the Board, and their elimination would seriously undermine the whole scheme, both technically and economically.

The proposed scheme would control a flood equal to that which occurred in 1894, the largest one on record. But it would not protect against all floods. The success of the scheme, therefore, would hinge in part upon the enactment and enforcement of floodplain zoning, aiming to control the increase in flood-loss potential. Unfortunately, the Fraser River Board had little to say about floodplain zoning, other than suggesting that 'the responsible authorities examine the use of lands in the floodplain with a view to restricting the use of those lands to developments that would suffer least from flooding'. The lack of a clear definition as to who the responsible authorities are inevitably means that the problem is unlikely to be studied! And without such a study regulations are unlikely to be forthcoming.

Reliance is placed on the local authorities to enact floodplain regulations. It seems, however, that such regulations will only be enacted and enforced if *all* local authorities do so. Without such uniform action individual local authorities will fear that they will lose industrial and urban development to other areas that do not enact regulations.

D. The Fraser River experience in a wider context

Adjustment to floods in the Lower Fraser Valley has been characterized by four main features. First, it has generally been a reaction to crisis, with short periods of feverish activity punctuated by longer intervals of inaction and apathy. The consequence has been a piecemeal approach which has been corrective rather than preventive in character. Second, there has been little innovation in flood policies over the years. Reliance has continued to be placed on a narrow range of adjustments – mainly flood protection, emergency action, and flood relief. The Fraser River Board considered other possible adjustments, such as floodplain zoning, but gave them only cursory attention, possibly reflecting the uncertainty as to which level of government would implement such measures but also reflecting a bias that characterizes most engineering reports, a bias towards adjustments that involve the construction of control works.

A third feature has been the increase in the share of costs of flood protection assumed by the general public. Today the senior governments pay more than half the costs of flood protection in the Valley, the remainder being borne by the local areas. A consequence of the shift of the financial burden has been to make individual floodplain occupants less concerned about the flood hazard. The extra costs of floodplain location are not brought home as vividly as they would be if he had to carry most of the cost. Moreover, the present policy of providing flood relief without any obligation on the part of the floodplain occupant to take remedial action has further reduced his concern about the problem. In a sense, the provision of such relief has furnished an incentive to further increases in the flood-loss potential.

A further feature has been the tendency to consider the flood problem in isolation from other problems of developing the Lower Fraser Valley. Consequently, while some government policies have tried to reduce potential flood losses, others have been encouraging developments in the floodplain which would increase such losses. The Water Resources Branch, for example, has been busy building and repairing dykes in an effort to curb losses, while the Department of Highways has been building highways through the floodplain, and the Department of Industrial Development has been promoting development in locations subject to flood hazard!

The Fraser River experience offers some lessons for policies relating to flood management elsewhere. First, it suggests that a concentration on a narrow range of alternatives may lead to inefficient adjustment. Sometimes, when the suggested course of action is blocked, no action is taken at all. Second, it shows that there are dangers in removing completely from the individual the incentive to consider the risks he is running by occupying the floodplain. The greatest danger is that activities will move into the area that might be more efficiently located elsewhere. To the extent that their flood losses are covered by public expenditures, their location in the floodplain is publicly subsidized. The adoption of such adjustments as floodplain zoning, flood proofing, and flood insurance would encourage the individual to consider the relative advantages of floodplain location versus location elsewhere, and at the same time reduce the potential burden on the public purse. Finally, it emphasizes that the lack of a clear definition as to who is responsible for dealing with flood problems generally means that there will be inaction. Floodplain dwellers will assume that the Government is dealing with the matter, while government agencies assume that it is beyond their terms of reference. Meanwhile the flood loss potential continues to grow. Catastrophe is the inevitable result.

REFERENCES

FRASER RIVER BOARD [1963], *Final Report on Flood Control and Hydro-Electric Power in the Fraser River Basin* (Queen's Printer, Victoria, British Columbia).

HOYT, W. A. and LANGBEIN, W. B. [1955], *Floods* (The Ronald Press, New York).

HUTCHINSON, B. [1950], *The Fraser* (Clark Irwin and Co., Toronto).

KATES, R. W. [1962], *Hazard and Choice Perception in Flood Plain Management*; University of Chicago, Department of Geography Research paper No. 78.

KRUTILLA, J. V. [1966], An economic approach to coping with flood damage; *Water Resources Research*, **2**, 183–90.

LEOPOLD, L. B. and MADDOCK, T. [1945], *The Flood Control Controversy* (The Ronald Press, New York).

MURPHY, F. C. [1958], *Regulating Flood Plain Development*; University of Chicago, Department of Geography Research Paper 56.

RENSHAW, E. F. [1961], The relationship between flood losses and flood control benefits; In White, G. F., Editor, *Papers on Flood Problems*, University of Chicago, Department of Geography Research Paper No. 70, pp. 21–45.

SCHAEFFER, J. R. [1960], *Flood Proofing: An Element in a Flood Damage Reduction Program;* University of Chicago, Department of Geography Research Paper No. 65.

SEWELL, W. R. D. [1965], *Water Management and Floods in the Fraser River Basin*; University of Chicago, Department of Geography Research Paper No. 100.

U.S. GOVERNMENT [1966], Executive Order No. 11296, August, 1966, U.S. 89th Congress Second Session, House of Representatives, Report of the Task Force on Federal Flood Control Policy, *House Document 465*, Washington, D.C.

U.S. SELECT COMMITTEE ON NATIONAL WATER RESOURCES [1959], *River Forecasting and Hydrometeorological Analysis*; Committee Print No. 25.

WHITE, G. F. [1939], Economic aspects of flood forecasting; *Transactions of the American Geophysical Union*, **20**, 218–33.

WHITE, G. F. *et al.* [1958], *Changes in Urban Occupance of Flood Plains in the United States*; University of Chicago, Department of Geography Research Paper No. 57.

WHITE, G. F. [1964], *Choice of Adjustment to Floods*; University of Chicago, Department of Geography Research Paper 93.

10.III. Human Responses to River Regimes

ROBERT P. BECKINSALE

School of Geography, Oxford University

Although flood control, irrigation, and the generation of water-power are carried on under a wide range of hydrological conditions, they are especially important where wide seasonal fluctuations of discharge occur. Here the period of high discharge poses particular flood-control problems, as well as providing sufficient water, when adequately impounded, to carry on irrigation and water-power generation during the dry season. It is important to recognize that few single-purpose projects are now being initiated, and that it is the multi-purpose schemes which represent the most sophisticated response by man to seasonal variations of discharge.

1. Flood control and streamflow routing

The need for flood control usually follows the settlement of people on the rich soils of a floodplain. The paddy cultivation of lowland monsoon Asia and the ancient irrigation systems of the Nile and Euphrates were dependent mainly upon river regimes with a violent warm-season flood. Records at the Roda gauge near Cairo date back to A.D. 640. Here and elsewhere little could then be done about exceptional low-water years, whereas high floods could be partly controlled by constructive levées. On the lower courses of many monsoon rivers (with AM; AW regimes) the natural levées were soon strengthened artificially and often resulted in a natural raising of the channel bed. Eventually over vast areas the rivers were flowing along ridges well above the adjacent floodplains. No doubt increasing deforestation and cultivation near the watersheds added to the silt content of the streams and to deposition in the lowland channels. Also the appreciable rise in sea-level since about 12,000 B.P. has greatly decreased the gradients of coastal plains.

Given similar regimes, the problems of flood control tend to be least on rivers with sufficient gradient to scour their beds and greatest on rivers incapable of preventing channel deposition. A river with a reasonable gradient can be controlled by dams, adequate storage reservoirs, dykes, and channel improvements. This is exemplified on a small scale by the Rhône just above Lake Geneva, where the river floods violently from snow and ice-melt in early summer and then spreads over a wide bed, much of which is exposed in the low-water season. In 1895 an attempt was made to improve the reach between Brigue and Lake Geneva by constructing eleven flood-dykes and several jetties to restrain the river to a narrower central channel. But the jetties proved to be wrongly sited, and the river-bed silted, causing the dykes to be overtopped and breached by floods. From 1928, after a scientific study, the hydraulic characteristics of the central channel were improved by making a dyked and graded bed. In many parts the river now began to scour, and only exceptional floods now overflow into the area between the channel dykes and a parallel line of external dykes. With the aid of drainage devices, the former floodplain has been transformed from marshes to fertile agricultural land (UNESCO, 1951a).

The floods at Florence in 1966 demonstrate the violence of Mediterranean (CS) river regimes. Owing to channel silting and artificial levée building, the Arno even in normal high water rises above the level of much of the adjacent city. Only extensive channel dredging, dyke strengthening, and an elaborate system of reservoir control near the watersheds will ensure that the riverine parts of the city are not flooded occasionally every few centuries.

On rivers with very low gradients and heavy silting, flood control today is largely a question of distant headstream control, coupled with storage reservoirs, levées or embankments, channel improvements, flood spillways, pumping stations, and flood forecasting. The problems of flood control have been overcome on a small scale on the Waal in the Netherlands, where the river regime is moderate. Here main dykes enclose a major (winter) bed 1 m to 2 km wide. The polders between these master dykes and the minor (summer) bed are protected against high water in summer by dykes only 3 or $3\frac{1}{2}$ m above the mean summer river-level. When the river floods most of the drainage is carried by the summer bed, but the winter bed acts as a storage reservoir for water in excess of the capacity of the central channel.

On a greater scale the lower Mississippi–Yazoo, with a floodplain of 75,000 km², demonstrates the problems of flood control of a meandering river with a high silt content and moderately violent regime. The natural levées are usually 3–4·5 m above the swamps 3–5 km back. Under the incentive of rich bottom land for cotton and corn, the levées have been strengthened artificially and now extend for more than 3,500 km and allow the channel flood-level to rise 7·5 m above that of the adjacent plain. Channel shortening by means of cut-offs, dyke strengthening, and the construction of flood spillways leading to areas where flooding can be controlled, have greatly lessened the loss of life and damage to property. In this protection, upstream water-control and an efficient flood-forecasting service also play an important part.

Flood control on low-gradient rivers with a large sediment load becomes most difficult where the regime is violent (AM; AW). Abnormal daily spates as well as high seasonal floods occur throughout monsoon Asia, where a dense population, largely dependent on paddy, needs summer floods but also requires protection against abnormally high waters. Here flood-control structures dominate the landscape. The North China plain of about 324,000 km², exclusive of East Shantung, is dominated by the levées which stretch for 720 km along the Hwang Ho and 140 km along the Yungting Ho. The bed of the former, or Yellow River, near Kaifeng lies at a maximum of 15 m above the plain. As its drainage basin includes

288,234 km² of loess and is liable to winter frost, its silt content is notoriously high and its bed is being steadily raised (UNESCO, 1951a, pp. 309–4). During the last 4,242 years it has changed its lower course completely 7 times, breached its dykes more than 1,170 times, and overflowed them about 425 times. The annual flood damage is large and involves more than 29,600 km². Among the

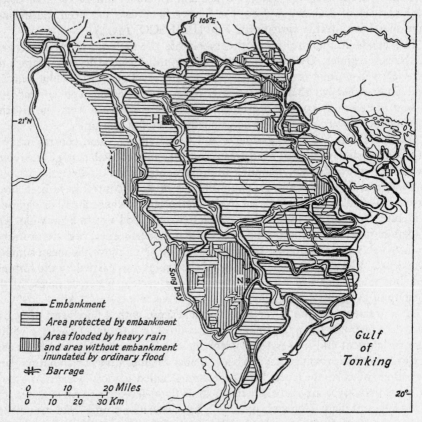

Fig. 10.III.1 Embankment system of the Red River delta, North Viet-Nam.
H. Hanoi   HP. Haiphong   N. Namdinh.
(Adapted from United Nations, 1966a, p. 15).

suggested remedies is a desilting basin to supply water free of coarse sediment (?silica particles) for irrigation. A great dam has been built across the river above Loyang, and for some years a scheme of 45 smaller dams and 26 reservoirs has been under construction higher upstream to regulate floods and reduce soil erosion.

Similar but smaller floodplains exist in most Asiatic countries, and only a few have as yet solved the flood-control problem. In North Viet-Nam the Red River and other streams form a delta that extends about 150 km inland (fig. 10.III.1). The Red River, which is 1,200 km long and rises in China, meanders for its last 220 km through a deltaic floodplain, sown almost entirely to paddy. Its annual discharge averages 3,900 m³/sec, but since recordings began has varied from 700 m³/sec to 35,000 m³/sec. The whole basin has a mean annual rainfall of 1,500 mm, and the lower course receives 1,800 mm. As the whole delta lies several metres below the highest flood-level, the people have fought a constant battle to contain floods in excess of those needed for paddy cultivation. In early times individual landowners strengthened their own dykes, but since the thirteenth century a central organization to protect the whole delta has evolved (United Nations, 1966a). This co-operative effort culminated between 1920 and 1944 in the strengthening of the complete dyke system. Levées now extend for 1,400 km and contain about 120 million m³ of earth. Only one serious breach has occurred since 1926.

Many of these basins in Monsoon Asia are so large and so fragmented politically that full flood control will need enormous capital and elaborate co-operation, and probably will not be realized for centuries. However, elsewhere many small basins and a few large basins (notably the Columbia, 772,000 km²) have already been controlled and instrumented sufficiently to allow scientific flood forecasting and streamflow routing (Rockwood, 1961; Lewis and Shoemaker, 1962). These ideas were first used comprehensively in 1943, and have progressed in value with the efficiency of the river control, with the knowledge of river hydraulics, and with the use of computers. Streamflow or flood routing provides a means of translating a hydrograph from an upstream to any downstream point so as to express the effects of the intervening channel and valley storage (Bruce and Clark, 1966, p. 197). It assumes that on a given reach of a river the outflow is equivalent to the inflow minus the amount of water stored in that reach. In practice, the technique of flood routing and forecasting is based on the assumption that:

1. rivers consist of reaches with relatively constant flow characteristics and that gauges or more detailed recordings exist at the entrance and exit of these reaches;
2. the routing period, or time interval used in the flood-routing formula, is the time taken for the flood wave to traverse the reach; and
3. the reaches are sufficiently short and self-contained to ensure that the storage in each is a function of the inflow and outflow of the river, or, in other words, is not unduly affected by local tributaries. If tributaries are present and are too small to warrant the establishment of a separate reach below their entrance their inflow should be added to that at the upstream end of the reach in question.

For flood routing the data required are the stage–discharge curve and the stage–storage curve for each reach of the river. The latter curve can be constructed from detailed flood recordings by reversing the routing procedure used. Thus, from past floods the difference between the inflow and outflow gives the amount of storage in the reach; the progressive sum of these storages plotted against the stage at the downstream end of the reach provides a stage–storage curve. It is necessary to begin the summation at a time when storage is about zero, that is when inflow and outflow of the reach are about equal. The methods and graphs are summarized by Bruce and Clark [1966, pp. 197–202]. Mechanical and electrical routing devices are installed on many rivers, and forecasts of flow and flood levels based on them are considered essential to make the optimum use of flow and to minimize flood damage. On uncontrolled rivers flood forecasts tend to be less accurate, but they still provide useful warnings, so that people and movable goods can be removed to safety and labour forces alerted. On partially controlled rivers reliable streamflow routing allows reservoir storage to be used most efficiently and economically. Thus flood control demands considerable accommodation space to be left inside reservoirs, whereas irrigation demands the maximum water storage. If a reservoir is kept too empty the irrigation may suffer from water shortage; if it is kept too full floods may damage the low-lying plains. Needless to say, the bigger the storage capacity, the better for all concerned, except those displaced by the reservoir. The phenomena of freeze and thaw which affects D regimes enters equally into streamflow routing. Scientific recordings and predictions now render obsolete the annual sweepstake at Yukon city, where a bell tied by a rope to a stake in the river ice gave the news and time of the break-up for the lucky winner. On tidal rivers, where for long distances the so-called tide is merely a freshwater wave moving upstream, the diurnal rise adds a further complication, as in East Pakistan, where over one-third of the total area (140,000 km^2) is liable to serious floods.

2. Irrigation

The application of water to the soil, or irrigation, is undertaken mainly to grow crops in dry climates and in humid climates with a marked dry season. It is, however, also widely practised in humid regions with a moderate rainfall in order to obtain higher yields and special crops. Thus, except in arid climates, irrigation is either supplemental or complementary to seasonal rainfall. The expenses involved are often offset by marked advantages of higher yields, better harvesting weather, and easier pest control. Also in some countries out-of-season products fetch extra prices on the national markets.

Most irrigation projects are on floodplains and valley floors, the main exceptions being where mountain springs are led overground to hillside terraces, as is common, for example, in Japan. Floodplains, if well drained, are often rich and easily cultivated; many are enriched almost annually by sediments; all are relatively flat and can be reached by simple water-lifting devices. In Eurasia the uncertainties and limitations of a seasonal flood with AM, AW, and CS river regimes led early to the building of earthen dams and to the invention of many water-lifting implements. On the Indus a wide variety of tanks and irrigation canals existed by about 3000 B.C.; on the Euphrates in Babylon earthen dams and an elaborate code of water laws existed in 2050 B.C. Most of the ancient methods of lifting water artificially are still in use. They were based on three elementary principles: the LEVER, as in the Egyptian *shadoof*, the Indian *picottah* and the Spanish *cigoñal*; the horizontal revolving WHEEL fitted with buckets round the rim, as in the Persian wheel, the Egyptian *sakiya*, Punjabian *harak*, and Spanish *noria*; and the SCREW, an invention of Archimedes of Syracuse in about 200 B.C. The last-named consists of a wooden cylinder containing a helix (corkscrew-shaped diaphragm) up which water is forced when the lower end is placed at a low angle in water and rotated.

Today irrigation is practised either seasonally or all the year, mainly where the river regimes are AM, AW, CS, HG, and HN. The seasonal form, often called basin irrigation, is commonest alongside uncontrolled rivers and is dependent on the flooding of embanked areas either direct or by means of artificially filled inundation canals. In contrast, perennial irrigation needs a partly controlled river with an almost constant water supply, thereby allowing two or three crops to be grown annually on the same plot. The basin method was, and still is, quite extensive, as many tropical and subtropical rivers rise 7 or 11 m in summer, and primitive lifting devices can raise the water farther to higher terraces. Thus in Egypt the Archimedes' screw raises water about 75 cm into a basin, whence a sequence of shadoofs, each with a rise of about 2·5 m, allows the upper terraces to be reached.

In some semi-arid regions with wadis that experienced a brief flow at one season (AW/BS regime) the peasants cultivated the wadi-floor after the rains. More commonly earthen dams were constructed across the valley to form tanks, such as survive in thousands in southern India and Ceylon. During floods much sediment was deposited in the reservoirs and in the floodplain. Recent analyses of the Nile 'mud' (UNESCO, 1951, pp. 279–99) show that the maximum concentration of silt in the river water reaches about 2·5 g per kg in August. If, as is usual, the basins are flooded to a depth of 1 m the layer of silt averages about 2·1 mm or about 10 tons per acre. Analyses of the chemical composition of the silt with regard to fertilizing properties revealed the following percentage (Table 10.III.1) of the chief plant nutrients and the approximate amount supplied per feddan or acre annually.

TABLE 10.III.1

	Percentage total deposition	Amount (kg)
Phosphorus (P_2O_5)	0·24	25·2
Potash (K_2O)	1·07	112·5
Lime (CaO)	4·16	430·6
Organic matter	2·42	254·1
Nitrogen (included in organic matter)	0·13	13·5

Thus the Nile mud is fairly rich in lime, potash, and phosphates and relatively deficient in nitrogen. However, the mechanical structure of the soil, when well drained, is very favourable to micro-organisms, especially nitrifying bacteria.

The main disadvantage of high-sediment content in rivers is the silting-up of reservoirs and of irrigation ditches. Probably the first dams with sluices near their base adequate to prevent excessive silting were the Periyar in southern India (1895) and the Aswan (1902). On the latter the sluices are kept open and flood-water allowed to pass through until the flood is lessening and its silt-content decreasing. The surest way of avoiding excessive sedimentation is to construct elaborate desilting basins, as in the Imperial valley schemes of the United States and at the new High Aswan dam. Without such desilting methods most of the world's great reservoirs on rivers will be largely silted-up within a few centuries, and many artificial reservoirs built in the twentieth century have already suffered appreciable sedimentation.

Irrigation practices and irrigated landscapes have evoked a large literature, including a global summary by Cantor [1967]. Of the numerous crops irrigated, paddy is supreme. It needs an almost constant water supply until towards harvest, and the more prolific varieties are usually grown where irrigation supplements a seasonal rainfall which helps to keep the basins filled and prevents problems of salinization.

Whereas Chinese, Japanese, and Arab (Moorish) irrigation methods dominated from early times in most of the Old World, in the New World the existing practices were greatly disrupted by the Spanish Conquest. In the United States irrigation did not revive until after about 1850. Here and elsewhere modern engineering techniques began to affect irrigation in the late nineteenth century. Up to 1902 the highest dam was at Furens in France, 170 ft (52 m). However, in the next two decades the increasing development of the petrol engine, of concrete construction, and of electrical equipment led to a great interest in dam building and water-pumping. By the 1940s dam erection was almost a national symbol of progress, especially as international concerns such as the World Bank made loans available for water-resource development. The colossal earth-shifting machines and cranes of some highly mechanized countries were offset in others by labour forces of up to 30,000 or 36,000 workers on a single dam. Richer nations vied and co-operated with each other to help and advise poorer nations in dam construction. During December 1962 there were at least 650 sizeable dams under construction, of which 217 were in the United States and 122 in Japan (United Nations, 1966b; Int. Comm., 1964; *World Almanac*, 1967, pp. 261–9). It is certain that the years 1930–70 will witness a greater increase in storage reservoirs and in irrigation acreage than has occurred in any previous century. Although many of the dams are multi-purpose and some are primarily for hydroelectric power, the post-1930 progress in water control for irrigation and other purposes may be judged on the growing size of barrages and reservoirs. In the 1930s the Hoover or Boulder Dam (221 m) was by far the tallest in the world. By the early 1970s it will be exceeded in height by at least thirteen others ranging up to 301 m (Inguri, U.S.S.R.). Most of these tall structures are of concrete, whereas the world's most massive dams are mainly earthfill or rockfill, usually with concrete sections. Of the twenty largest dams only Fort Peck, U.S.A. (125 million cubic yards; built in 1940) will pre-date 1950. Similarly, of the eleven huge barrages over 5 km long, only Fort Peck (6½ km) is more than thirty years old, and the barrage under construction at Kiev will be 41 km in length. The world's greatest man-made lakes have become truly impressive. Whereas Lake Mead (Hoover Dam) led the way in 1936 with about 31 million acre-feet of water, by the early 1970s it will be surpassed by at least eleven reservoirs, irrespective of three other dams which already greatly increase the volume of natural lakes. The details in acre-feet are as follows: Kariba, Rhodesia, 130 million; Sadd-el-Aali (High Aswan), 127·3 million; Akosombo, Ghana, 120 million; Manicougan, No. 5, Canada, 115 million; Portage Mountain, Canada, 62 million; Krasnoyarsk, U.S.S.R., 59·4 million; Volga, V.I. Lenin, 47 million; Tabaga, Syria, 31·6 million. There are also eight other artificial reservoirs with a capacity of between 20 million and 28 million acre-feet. Such vast reservoirs allow interesting new possibilities of long-term storage.

No certain means exist of predicting at the start of a hydrological year the discharge for the coming year. The best that can be done from a long period of records, say of fifty or sixty years or more, is to find the mean annual trend, which varies appreciably with different decades (Moran, 1959; Hurst, Black, and Simaika, 1965 and 1966). However, the really large reservoir allows storage to meet most deficiencies. Thus the High Aswan or Sadd-el-Aali reservoir has four functions; to protect against dangerous floods; to produce hydroelectricity; to store sufficient water for use during the low-water stage; and to reserve excess water in abundant years to augment the supply in lean years. The total capacity at the maximum level of R.L. 182 m. is 157·4 milliard m³ (1 million m³ a day = 0·36 milliard m³ a year). This is allotted as follows:

To silt trap	30 milliards	Reservoir Level 146 m
Over-year storage	90 milliards	Up to R.L. 175 m
To flood protection and annual storage	37·4 milliards	R.L. 182 m

As the silt-trap content remains dead storage, the over-year storage will be confined between R.L. 146 m and R.L. 175 m (content 120 milliards). The theory of this long-term storage is discussed by Hurst, Black, and Simaika [1965 and 1966]. The biblical seven fat years and seven lean years are now a legend; the only casualties may well be the marine life that thrived at the delta mouths.

With such storage expansion and river control it is not surprising that the irrigated acreage in the world has increased rapidly in the twentieth century. However, precise details are hard to obtain, especially as it is difficult to decide when the supplemental watering of crops causes them to be grouped under irrigation (Cantor, 1967; FAO, 1967, Table 2; Highsmith, 1965). In the late 1930s probably over 200 million acres were irrigated in the whole world. By the early 1960s this was estimated by some authors to have increased to about 370 million acres or 13% of the world's cultivated area. But, as Table 10.III.2 shows, the twenty-four leading countries for irrigation alone then had about 440 million acres irrigated, and if these returns were reasonably accurate the world total would be nearer 470 million acres. Irrespective of the accuracy of the global total, over 80% of all irrigated lands are in non-Soviet Asia, about 7·5% in the United States, and 5% in the U.S.S.R. The probable acreage under or available for irrigated crops in the early 1970s is also estimated in the Table. During the 1970s, if scheduled projects progress as expected, the world total under irrigation might well exceed 500 million acres. Remarkably little of this expansion will be in truly arid regions.

TABLE 10.III.2 Irrigable or irrigated land (millions of acres)

Country	Early 1960s	Early 1970s (estimated)*
Chinese Republic	183	190+
India	65	75
United States	35	37·5
U.S.S.R.	23·5	35
Pakistan	26·6	30
Indonesia	15·9	16
Iran	11·5	12
Iraq	9·1	9·5
Mexico	8·7	11
Japan	7·7	9
Italy	6·9	7·7
Egypt	6·1	7·5
France	6·2	6·5
Spain	4·6	5·8
Thailand	4·0	5·5
Turkey	3·2	4·5
Korea (Republic)	3·0	3·1
Argentina	2·8	3·8
Peru	2·8	3·2
Chile	2·7	3·1
Nepal	2·7	2·9
Australia	2·0	3·3
Sudan	2·0	3·0
Philippines	2·0	2·2

* Based on projects already planned or under construction. They may be rather conservative, for example, for Argentina.

By far the leading country for irrigation is the Chinese People's Republic, where large river-control schemes, much construction of hillside terracing, and rapid progress in mechanization, including electric pumping, have been, and still are being, undertaken. The reliability of the irrigation acreage given in the Table may be checked roughly by comparison with the area under paddy (for the most part irrigated). In 1966 the Chinese mainland had about 220 million acres under paddy.

The Indian subcontinent has well over 100 million acres under irrigated crops (cf. 115 million acres under paddy in 1965). Probably nearly a quarter of this is watered by wells or underground aquifers and, strictly speaking, should not be included in a discussion of the use of open channels. Here paddy (with jute in East Pakistan), millet, cotton, and corn are the leading hot season (*kharif*) crops, and wheat and vegetables the main cool season (*rabi*) crops. Several large new projects are under construction in the north-west, and a large scheme on the Mahanadi delta in Orissa. The Indus is probably the world's greatest river for irrigation. India and Pakistan have agreed upon a water-control scheme for it, and a large new dam has been built under this agreement. In Pakistan alone the river irrigates about 30 million acres, and the huge projects include the Sukkur barrage (completed in 1932), with 36,000 miles of main channels and distributaries supplying water to 2·6 million hectares (7·5 million acres); and, farther downstream, the Ghulam Mohammed barrage (headworks completed in 1955) that irrigates about 1,112,000 hectares (2·75 million acres), of which 40% is perennial (fig. 10.III.2).

In the United States, the third leading country for irrigation, the projects are mainly in the seventeen western states, the chief being California, with about 7·5 million acres. The crops include those of high dietary appeal or value, such as fruits and vegetables, for which there is an all-the-year-round demand by a nation endowed with a superabundance of basal foodstuffs. Cotton also is important, because of freedom from boll-weevil infestation and ease of machine picking. Fodder crops also play a large role in supplying safety and fattening for extensive grazing grounds (U.S. Dept. Agr., 1962). In 1939, when about 18 million acres were irrigated, the water withdrawal was about 63 million acre-feet, of which 83% came from surface streams. In 1959 the 34 million acres irrigated used 103 million acre-feet of water, of which only about 56% was derived from surface flow. The great modern interbasin water exchanges, especially on the Columbia and in California, have restored a wider and greater use of river water.

The fourth country for irrigation is the U.S.S.R., where large developments have progressed in the last thirty years, particularly in Soviet Central Asia near the Aral Sea. Here before 1960 the irrigators used mainly seasonal water from the smaller rivers fed by snow and rain (HN regime) and rising to flood stage in spring. Subsequently great projects have involved the larger rivers, the Amu-Dar'ya (or ancient Oxus) and Syr Dar'ya, which are fed largely by mountain snow and ice and maintain a high flood throughout the summer (HG and HN regime). It was calculated (Lewis, 1962; Vendrov *et al.*, 1964) that about 5 million hectares (12 million acres) could be irrigated here by 1965 and another 15 million hectares (37 million acres) by 1985. However, salinity and rapid seepage from distribution canals have been serious problems, and probably 'at the present level of efficiency, roughly 22 million acres is the ultimate amount of land that could be irrigated' in Soviet Central Asia (Lewis, 1962, p. 114).

Most of the countries with less than 20 million acres under irrigation have made considerable extensions since 1950. In Mexico, where cotton and sugar cane are important crops, modern methods have allowed some rehabilitation of older irrigated tracts rendered useless by excessive salinization. In Egypt the modern schemes include the Sadd-el-Aali dam, which is about 364 ft high,

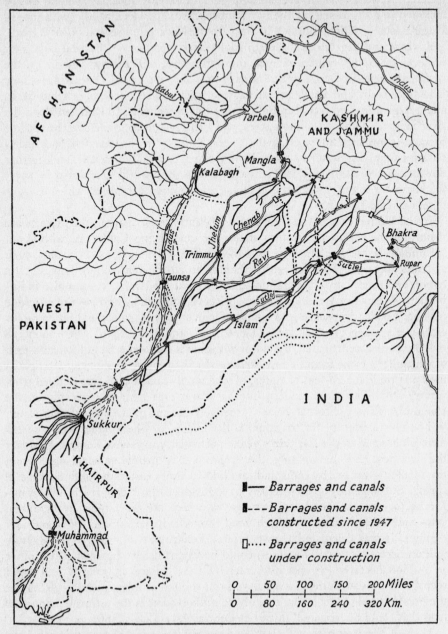

Fig. 10.III.2 Water Resource Development Projects in the Indus Basin in 1967.

The irrigated area extends within a few miles of the canal system. The territory was partitioned politically in 1947 and the international Indus Water Treaty was signed in 1960. The international boundaries are shown approximately by a pecked-dotted line. (Adapted from United Nations, 1966a, pp. 56 and 61.)

11,480 ft long, and holds back a reservoir 242 miles in length and 127 million acre-feet in capacity. By the early 1970s it will irrigate an extra 2 million acres and convert a further 700,000 acres from basin to perennial irrigation. The Sudanese Republic, with its large unused irrigable area, will benefit indirectly from the above scheme, as under reallocations agreed upon with Egypt in 1959, the Sudan Manaquil project of the Gezira receives 25% of the Nile water plus a margin for evaporation. In return the Egyptian (U.A.R.) Government acquired an eighteen years' lease of unused Sudanese water and permission to flood the Wadi Halfa valley with water held up by the Sadd-el-Aali barrage. In addition, the Sudan administration has a large new project under construction (132 miles upstream from the Sennar Dam on the Blue Nile) that was made possible by the Nile Waters Agreement and a loan of $32 million from the World Bank.

3. Water-power

From early times water-power was used to drive revolving wheels, such as *norias* for irrigation and, by means of shafts, stones for grinding cereals and pulses. In 1086 in the parts of England recorded in Domesday Survey there were at least 5,624 grist mills, many no doubt having horizontal wheels. During the twelfth and thirteenth centuries a minor industrial revolution was caused in some textile districts of western Europe by the application of water-power to drive fulling stocks by means of undershot wheels. Gradually where streams could be easily dammed or diverted the undershot wheel gave way to the more efficient overshot wheel, which consisted of a series of cups filled from above and forced downward by the weight of a relatively small amount of water. By the end of the eighteenth century in Europe and the United States quite elaborate machinery was water-driven, and the so-called Industrial Revolution in both textiles and iron-processing and manufacturing was based largely on water-power. In spite of the growing use of steam and later of internal-combustion engines, water-power remained significant in some areas well into the twentieth century and, for saw-mills and grist, is still used in a small way in some districts.

The introduction of electricity revolutionized the uses of water-power and led rapidly to the conversion of traditional wheels to electricity production. The first such conversion for industrial purposes was probably that undertaken by Aristide Bergès in 1869 in his woodpulp factory at Lancey in the Dauphiné Alps. The first hydroelectric station in France (of 900 h.p. on the Valserine), and the first commercial generating thermal stations in New York and London were built in 1882. Two years later Marcel Deprez began to transmit electrical current over appreciable distances from hydroelectric stations in France. Within a decade several other countries had small distribution networks. The twentieth century brought a tremendous expansion, particularly after 1918, when 24-hour working of plant became more general, and after 1945, when dam-building became extraordinarily popular. This expansion may be illustrated from Canada, where the hydroelectric generating capacity grew from 0·15 million kW in 1900 to 4·6 million in 1930, 7·8 million in 1949, 13 million in 1955, and 21·7 million kW in 1965 (Davis, 1957).

TABLE 10.III.3 Hydroelectricity production (million kWh) and installed capacity (in 1,000 kW) of leading producers

Country	Production 1965	1966	Hydro % of total electricity output (1965)	Installed hydro capacity
United States	197,001		17	44,492
Canada	117,063		81	21,711
U.S.S.R.	81,431		16	22,244
Japan	76,739		40	16,279
Norway	48,858		99·8	9,783
France	46,429	50,736	49*	12,683
Sweden	46,423		95	9,278
Italy	42,367	44,000	51*	13,955†
Brazil	25,515		85	5,391
Switzerland	24,015		98	8,120
Spain	19,550		62	8,141
Austria	16,083	17,327	73*	4,054
W. Germany	15,365	16,800	9*	4,072
India	14,807		41	3,331†
Finland	9,488	10,516	67*	1,857†
Yugoslavia	8,985		58	2,265
Mexico	8,609		50	2,327
New Zealand	8,588		81	1,910
Australia	8,367		23	2,092
United Kingdom	4,625		2	1,760
Czechoslovakia	4,456		13	1,540
Portugal	3,983		86	1,311†
Chile	3,954		64·5	710
S. Rhodesia	3,864		94	705
Colombia	3,218		63	793†
Peru	2,625		68	680
Turkey	2,167		44	510
Bulgaria	2,001		20	768

* For year 1966.　　　　　　　† For year 1964.

In progressive societies living in countries deficient in coal and oil, hydroelectricity became the main motive force for factories as well as for domestic appliances and traction. Installation was cheapest where natural waterfalls occurred, as in glaciated uplands, where hanging valleys, ungraded rivers, and storage lakes abound. Thus in Norway about 60% of the main hydro sites have natural falls with drops ranging from 300 up to 1,008 m, while in Sweden and Finland, where the falls are lower, large natural reservoirs occur. Here and in many other mountainous and coal-deficient countries hydroelectricity has become the main source of power, and allowed the creation of manufacturing societies with a very high standard of living. In 1950 the percentage of total national electricity output provided by hydrogenerators was nearly 100 in Norway, 98 in Switzerland, 97 in Canada, 95 in Sweden and New Zealand, 88 in Italy, and 49 in France. On the other hand, in countries rich in other energy sources or lacking in waterfalls the hydro percentage of total national electricity output was 26 in the United States, 19 in West Germany, 3 in Great Britain, and zero in Denmark and the Netherlands. The situation in 1965 for all countries producing more than 2,000 million kWh of hydroelectricity annually is shown in Table 10.III.3 (United Nations, 1967, Tables 143 and 144).

Especially since 1918 electro-process industries have turned increasingly to large-scale hydro sources, such as the Niagara Falls. Certain industries are exceptionally greedy of power. Thus even in Canada, where large stations are favourable to cheap output, in the early 1950s the cost of energy was equal to nearly 11% of the value added by manufacture in paper products, 12·6% in non-ferrous metal products, and 14·6% in non-metallic mineral products. In the manufacture of newsprint the cost of energy amounted to 18% of the net value of the production; for aluminium and dissolving pulp the percentage was 27 and 30 respectively. Energy costs such as these attracted certain industries to hydro sites, and although their drawing power has weakened slightly today, they remain formidable, particularly for aluminium. In several countries two-thirds of the hydroelectric output is still used in the refining of non-ferrous metals, electro-chemicals, and pulp and paper manufacturing.

The world demand for electricity has grown so fast that it could be met only by greatly increasing the installed capacity of both thermal and hydro stations aided by advances in generating efficiency and in transmission. As modern hydroelectric plants already have a very high overall engineering efficiency, the future hope of improving the output per installed kW and the amount and cost per unit delivered to the customer lies in three methods.

First, in seeing that water flow or storage is adequate for capacity use all the year, especially where the regimes have a seasonal drop in discharge.

Second, in improving transmission. The long-distance bulk transmission of power at high voltages of 500 and 735–765 kV is being developed, especially in Canada, the United States, and Sweden. In the U.S.S.R. in 1963 the Bratsk hydro station on the Angara River in Siberia was brought up to 3·6 million kW installed capacity and was linked by a 500-kV line with Krasnoyarsk and the West Siberian power grids.

Third, in lowering the fixed costs, that is of initial constructions and installations. A large installation tends to be much cheaper than several smaller stations of an equivalent total capacity. Hence the tendency today to construct large storage reservoirs and/or long penstocks and tunnels. In Sweden, for example, it is common to utilize the head of a long stretch of rapids by building a reservoir and leading tunnels from it to a single large underground generating station. Since 1940 numerous investigations have been made into cheaper methods of dam construction and equipment installation, and certain companies and consortia have acquired a tremendous knowledge and experience (United Nations, 1957a). But the fixed costs remain high. In some of the huge Soviet schemes they

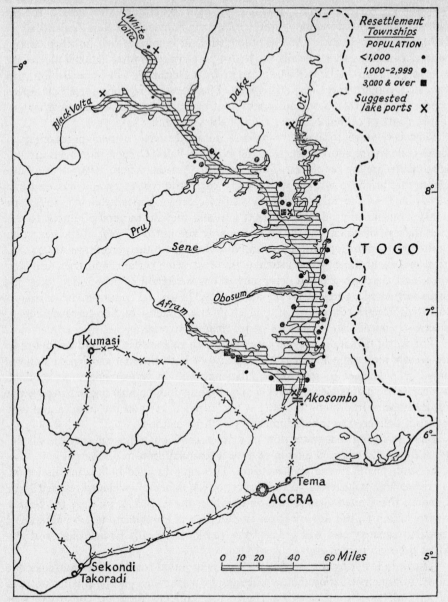

Fig. 10.III.3 The Akosombo Dam and the Volta River Project, Ghana, showing the main transmission lines (Adapted from Hilton, 1966).

have ranged from 12 up to 20% of the total outlay; in the small schemes of the Scottish Highlands the capital investment per hydro kW is two to four times that for conventional thermal generation; in Canada the large hydro units make the costs rather less unfavourable for water-power. However, the fact is that in many areas hydro power is either the only feasible local development or its high initial outlay is in the long run offset by benefits to irrigation, land drainage, and flood control. At the Sadd-el-Aali barrage about 210,000 kW are available for eight months and 50,000 kW for the flood months. The electricity here will be used for electro-ferrous and electro-chemical (fertilizer) processes as well as for pumping and general purposes. The large Kariba barrage (Reeve, 1960) on the Zambezi between Rhodesia and Zambia has an installed capacity of 705,000 kW and an output (in 1965) of 3,864 million kWh, which is almost equal to that of several other countries with double its installed power. A similar large project, also built by an Italian consortium, has just been completed in Ghana, across a gorge of the Volta River at Akosombo (fig. 10.III.3). The financial loans came mainly from United States' combines (Volta Aluminium Company), which guaranteed to buy power in bulk for thirty years and in return were granted freedom from expropriation and exemption from import duties on equipment. The power is especially designed for alumina reduction, but will also be used generally throughout southern Ghana, and is expected to cover costs and eventually to make a profit. The installed capacity is about 500,000 kW and can be increased to 768,000 kW. As fig. 10.III.3 shows, the large lake will provide fishing grounds, a 250-mile waterway, and irrigation for about 650 square miles around its shores. A considerable resettlement of the inhabitants will be necessary, but the total benefits are expected to offset this disruption and the inundation of an area of poor savanna covering about 3% of the national territory (Hilton, 1966).

The third method of cheapening the costs and efficiency of hydrostations is, where several different sources of energy exist in a region, to integrate all sources in an interconnected super-grid. This allows the economic exchange of power; mutual assistance in emergencies; the use of units with a low 'firm' capacity; the construction of giant generating stations; and the more efficient use of installed power by lessening the capacity kept in reserve (UNESCO, 1951b, pp. 224–54; OECD, 1961; Fed. Power Comm, 1965). Today in many countries hydro and thermal plants are considered complementary rather than rivals. As hydro-electricity can be rapidly turned on for addition to a transmission grid, even in areas where thermal plants predominate, it is becoming common to use surplus power of off-peak periods to pump water to storage reservoirs which can be returned to the grid as hydroelectricity at peak-demand hours. Many large regional, national, and international networks already exist. For example, in the United States the North West Power Pool and the Pacific South West Power Exchange ensure an efficient use of the great hydrosites on the Columbia and the Colorado. 'Power will be transmitted up to 870 miles . . . as far as southern California, at both 500-kilovolt alternating current and 750-kilovolt direct current' (Fed. Power Comm., 1964, p. 35). In the State of Washington the fifty largest hydrostations have a total installed capacity of over 10 million kW, and the Grand Coulée alone has 1,974,000 kW installed and a potential of 5,574,000 kW. The advantage of such large units are obvious when compared, for example, with those in the Scottish Highlands, where, owing to the small drainage basins, the building of more than fifty 'main' hydrostations has resulted in a total installed capacity of only 1,047,000 kW. In the latter area, a glaciated upland of 21,000 square miles and about 1 million inhabitants, the North of Scotland Hydro Electric Board was pledged to aid social and economic development generally. By 1966 it had supplied 96% of the premises, many of them remote and isolated dwellings, with electricity, but by then the total capital expenditure had risen to over £184 million. Thus the cost of firm power was several times that in countries with larger and more concentrated hydro resources (H.M.S.O.).

Sweden and France provide excellent examples of national integration schemes. In Sweden the bulk of the hydroelectricity is sent several hundred miles to the south, while in France the electricity production is predominantly

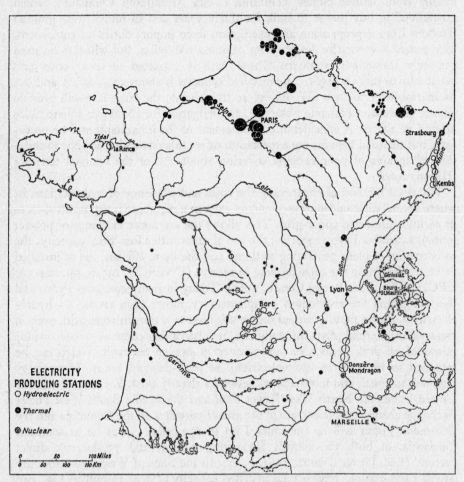

Fig. 10.III.4 Electricity Generation in France in the early 1960s, showing main thermal and hydrosites.

The circles are approximately proportional to the annual output, and may be judged from Kembs 840 million kWh and Donzère–Mondragon 1,890 million kWh. Small circles denote under 100 million kWh.

hydro in the south and thermal in the north (fig. 10.III.4). Still larger integration on an international scale is becoming common. All the Scandinavian countries are linked by transmission lines, and that between Sweden and Denmark, today of five cables, dates back forty years. In eastern Europe (Kish, 1968) the unified energy system or power grid of COMECON (the economic co-operation of the Soviet bloc in East Europe) has a total installed capacity of 35,000 MW.

Special interest in the world today centres on multiple-purpose dams, but irrespective of the variety of purposes, it is usually hydroelectricity that is the money earner, at least until other projects, such as elaborate irrigation, come into full operation. Many countries today finance construction schemes by loans from national and international agencies such as the World Bank, the International Monetary Fund, and the Inter-American Development Bank. This partly explains the unparalleled growth since 1950 of the number and global distribution of large hydrostations. Up to about 1950 the Hoover Dam (1,345,000

TABLE 10.III.4 World's largest hydrostations

Name	Capacity in 1,000 kW Installed	Ultimate	Year of initial operation
Sayansk, U.S.S.R.		6,000	U.C.
Krasnoyarsk, U.S.S.R.		6,000	1967
Churchill Falls, Canada		6,000	U.C.
Grand Coulée, U.S.	1,974	5,574	1941
Sukhovo, U.S.S.R.		4,500	U.C.
Bratsk, U.S.S.R.	3,600	3,600	1961
Solteira Island, Brazil		3,200	1969
John Day, U.S.A.		2,700	1968
Nurek, U.S.A.		2,700	1970
Volgograd (22nd Congress) U.S.S.R.	2,543	2,560	1958
Kubyshev (V.I. Lenin), U.S.S.R.	2,100	2,300	1955
Portage Mountain, Canada		2,300	U.C.
Iron Gates, Romania-Yugoslavia		2,160	U.C.
Sadd-el-Aali, U.A.R.		2,100	1967
Mica, Canada		2,000	1975
Robert Moses-Niagara, U.S.A.	1,954	1,954	1961
St. Lawrence Power Dam, U.S.A.–Canada	1,880	1,880	1958
Guri, Venezuela		1,757	U.C.
The Dalles, U.S.A.	1,119	1,743	1957
Chief Joseph, U.S.A.	1,024	1,728	1956
Kemano, Canada	835	1,670	1954
Beauharnois, Canada	1,586	1,641	1951
Ingari, U.S.S.R.		1,600	1970
Kariba, Rhodesia	705	1,400	1959
Jupia, Brazil		1,400	1966
Sir Adam Beck, No. 2, Canada	900	1,370	1954
Hoover, U.S.A.	1,345	1,345	1936

kW) and the Grand Coulée Dam (1,947,000 kW installed) were unrivalled in magnitude. Since then at least twenty-five other stations have been started or completed with an ultimate capacity of over 1,345,000 kW and another fifteen of between 1 million and 1¼ million kW. In Table 10.III.4 blanks under the installed capacity indicate that the statistics were not available; U.C. denotes under construction (United Nations, 1966b).

The increasing demand for electricity makes the unused potential hydro power of considerable significance. Table 10.III.5 shows that enormous amounts of unharnessed water-power exist in Africa, South America, and Asia. Even in North America and in Europe, except in a few areas, the unused potential is large (Young, 1955; Hubbert, 1962, p. 99).

TABLE 10.III.5 Estimated world's water-power capacity (In 10³ MW)

Region	Potential	% world total	Developed (1961)	% potential
Africa	780	27	2	0·3
South America	577	20	5	0·9
U.S.S.R., China, and Satellites	466	16	16	3·5
South East Asia	455	16	2	0·4
North America	313	11	59	19·0
Western Europe	158	6	47	30·0
Australasia	45	2	2	4·4
Far East	42	1	19	45·0
Middle East	21	1	—	—
WORLD	2,857	100	152	5·3

The future situation for water-power is largely a question of the cost per unit compared with that of thermal power sources, of the availability of capital for the initial installation, and of the possibilities of regional long-distance power exchanges. But hydrogenerators retain certain important advantages. They are non-pollutive, both of river water and of the atmosphere; they do not raise the water temperature; they have a relatively long life and are fairly easily maintained. In the United States the installed hydro capacity is expected to rise from about 40 million kW in 1963 to 80 million by 1981, but at the same time the proportion that water-power supplies to the total national electricity production will decline from 18 to about 15%. On the other hand, in many other parts of the world hydroelectricity will supply a steadily increasing percentage of the energy output.

4. Regional and inter-regional water-resource schemes

The majority of large water-resource schemes and dams are today multi-purpose and aim at controlling river flow for the best advantage of the economy of a drainage basin. The earliest comprehensive schemes were on the Rhône and

Tennessee. In 1921 the French Parliament laid down the principles of developing the Rhône for water-power, navigation, and irrigation. The expenses were to be met by the sale of electricity. But before the scheme got under way the Boulder Canyon Project Act in the United States (passed in 1928 and working under a multiple-purpose Federal Water Power Act of 1920) included in its terms of reference hydroelectricity, navigation, irrigation, flood and sediment control, and water supply for southern California. Here, too, electricity was the money earner, and indeed soon made the scheme viable. The trend towards multiple-purpose planning in the United States culminated in 1933 in the creation of the Tennessee Valley Authority, which soon produced a most highly developed drainage-basin scheme. Here twenty-eight multiple-purpose reservoirs are operated as a unit and the whole basin from watershed downward is under reasonable control (Patchett, 1943; T.V.A., 1963). A start had been made on this scheme when in 1934 the French *Compagnie Nationale du Rhône* was founded. The C.N.R. was a form of limited-liability company, with the Government acting as security for loans floated with the public. It lacked the broad powers and speed of decision enjoyed by the T.V.A., which consequently achieved its aims much more quickly. However, the C.N.R. scheme is nearing completion, and it and the T.V.A. remain the prototypes of similar and larger schemes elsewhere. The annual operation and maintenance costs and the benefit–cost relationships show that integrated basin development with large multi-purpose dams is in the long run a profitable investment that no riverine societies can afford to be without. By 1968 numerous countries were planning, or constructing, or had completed river-basin development schemes (United Nations, 1955–61, 1957b, 1958, 1960, 1962, and 1966a; White, 1962 and 1963). Among the many examples are the Damodar Valley Project in India; the Kitakami River Basin Project in Japan; and the Helmand Valley Authority in Afghanistan. The last-named includes most of southern Afghanistan and involves the reclamation and irrigation of a large area of desert and the provision of 'educational institutions, sanitation and public health centres, modern housing, resettlement, particularly for the nomadic tribes, cottage and small industries etc. in order to raise the levels of living of the people in general' (U.N., 1955–61, *Part 2D*, 1961, p. 8). The United States has since 1949 financed the building of eight dams on the Helmand River for hydroelectricity and irrigation. When finished the plan will restore to the area the rich agriculture that flourished centuries ago before over-cultivation and soil erosion turned it into desert.

These successful integrated socio-economic schemes have been for basins within a single national territory, whereas the basins of most great rivers, except in parts of the U.S.S.R., extend into several different political states, so that integrated basin projects need international agreements. The problem of international basin law is discussed on pp. 351–4. Here it may be noticed that development is much more difficult when only part of a basin can be controlled. The great schemes of ECAFE (Economic Commission for Asia and the Far East) began with flood control in 1949, and by 1952 had changed to multi-purpose river-basin development of water resources. The Lower Mekong River Project applies only to the lower course of a river which rises in China and drains 810,000 km² compared with the 99,000 km² of the Rhône. Yet even this restricted project involves national ownership by Thailand, Laos, South Viet-Nam, and Cambodia. Already over $100 million have been spent here on two large dams in Thailand and other improvements by twenty-one Western countries and the four riparian states. How successful international co-operation in drainage-basin development can be is also seen in the Columbia River agreement between the United States and Canada (U.S. Bur. Rec., 1941; Sewell, 1966), and the Indus agreement between India and West Pakistan. How unfortunate disagreement can be is seen in the Jordan basin, where Israel and the state of Jordan have constructed rival irrigation schemes which are extremely costly and mutually harmful (Smith, 1966).

5. Inter-basin water transferences

The twentieth century has popularized the technique of transferring water on a large scale from one drainage basin to another, with tremendous effects on the socio-economic geography at least of the receiving basin. In California about 70% of the water supplies are in the north, whereas 77% of the water needs are in the southern two-thirds of the state. Under the State Water Plan a 444-mile aqueduct leads water from the Sacramento–San Joaquin delta to the central and southern districts. More exciting is the transference of water across or through high watersheds, such as from the Colorado to Los Angeles. Frank Quinn (1968) has shown that the American West, which in the nineteenth century was won largely by the adjustment of rural communities to limitations of local water supply, now receives enormous quantities of water by interbasin transfers. 'One out of every five persons in the Western states is served by a water-supply system that imports from a source a hundred miles or more away. In total tonnage the amount exceeds that carried by all the region's railroads, trucks, and barge lines combined.'

Transference between basins draining to different oceans is becoming increasingly common. In the Andes at least one scheme in Chile and two schemes in Peru achieve a transference from the Atlantic to the Pacific. In New South Wales, Australia, the Snowy Mountains scheme leads water from the eastward Pacific slope to the westward Indian Ocean slope by means of two systems of tunnels. When completed in the early 1970s the nine power stations and seventeen dams will provide about 4 million kW at peak loads and about 2 million acre-feet of water, or enough to irrigate 640,000 acres of dry land in the Murrumbidgee–Murray basins.

Most of the water involved in these interbasin transference schemes is reservoir storage that would have run to waste in time of flood, and do not adversely affect the streams being used as feeders. But in at least one case, the Paraiba do Sul near Rio de Janeiro, a large water diversion (to feed a hydroelectric station outside the basin) has led to a serious loss of water in its lower channel.

Interbasin and inter-regional water transferences are in their infancy. Vast amounts of unharnessed hydroelectric power can in some countries probably be best used for pumping water long distances. Many nations have not yet realized what a desirable product water is to sell and what markets exist for distant pipeline distribution. Water purchasers always have an assured exchange, at least in agricultural crops. The conjunction of Congo and Sahara offers more possibilities to mankind than journeys to the moon, and is altogether a much simpler and less costly project.

REFERENCES

BRUCE, J. P. and CLARK, R. H. [1966], *Introduction to Hydrometeorology* (Pergamon Press, London), 319 p.

CANTOR, L. M. [1967], *A World Geography of Irrigation* (Edinburgh), 252 p.

DAVIS, J. [1957], *Royal Commission on Canada's Economic Prospects: Canadian Energy Prospects* (Ottawa).

F.A.O. [1967], *Production Yearbook 1966*; Vol. 20 (Rome).

FEDERAL POWER COMMISSION (UNITED STATES) [1965], *44th Annual Report* (Washington, D.C.).

HIGHSMITH, R. M. [1965], Irrigated lands of the world; *Geographical Review*, 55, 384-7.

HILTON, T. E. [1966], The Akosombo Dam and the Volta River Project; *Geography*, 51, 251-4.

H.M.S.O., *Electricity in Scotland* and *North of Scotland Hydro-Electric Board Annual Reports* (Edinburgh).

HUBBERT, M. K., Editor [1962], *Energy Resources*; National Research Council Publication 1000-D, National Academy of Science, Washington, D.C.

HURST, H. E., BLACK, R. P., and SIMAIKA, Y. M. [1965], *Long-Term Storage: An Experimental Study* (Constable, London).

HURST, H. E., BLACK, R. P., and SIMAIKA, Y. M. [1966], *The Major Nile Projects: The Nile Basin;* Vol. 10 (Cairo, U.A.R.) (see pp. 54-156).

INTERNATIONAL COMMISSION ON LARGE DAMS [1964], *World Register of Dams.*

KISH, G. [1968], Eastern Europe's power grid; *Geographical Review*, 58, 137-40.

LEWIS, D. J. and SHOEMAKER, L. A. [1962], Hydro-system power analysis by digital computer; *Proceedings of the American Society of Civil Engineers, Journal of the Hydraulics Division*, 88, 113-30.

LEWIS, R. A. [1962], The irrigation potential of Soviet Central Asia; *Annals of the Association of American Geographers*, 52, 99-114.

MICHEL, A. A. [1967], *The Indus Rivers* (Yale).

MORAN, P. A. P. [1959], *The Theory of Storage* (London), 110 p.

O.E.C.D. [1961], *Power System Operation in the U.S.A.*

PRITCHETT, C. H. [1943], *The Tennessee Valley Authority* (New York).

QUINN, F. [1968], Water transfers: Must the American West be won again?; *Geographical Review*, 58, 108-16.

REEVE, W. H. [1960], The Kariba Dam; *Geographical Journal*, 126, 140-6.

ROCKWOOD, D. M. [1961], Columbia Basin streamflow routing by computer; *Transactions of the American Society of Civil Engineers*, 126, 32-56.

SEWELL, W. R. D. [1966], The Columbia River treaty; *Canadian Geographer*, 10, 145-56.

SMITH, C. G. [1966], The disputed waters of the Jordan; *Transactions of the Institute of British Geographers*, No. 40, 111–28.

T. V. A. [1963], *Nature's Constant Gift: A Report on the Water Resources of the Tennessee Valley* (Knoxville, Tennessee).

UNITED NATIONS [1955–61], *Multiple-Purpose River Basin Development* (ECAFE). (Parts 1, 2A, 2B, 2C, 2D being Flood Control Series Nos. 7, 8, 11, 14, 18.)

UNITED NATIONS [1957a], *Bibliographical Index of Works Published on Hydro-Electric Plant Construction* (Geneva).

UNITED NATIONS [1957b], *Development of Water Resources in the Lower Mekong Basin*; Flood Control Series No. 12 (New York).

UNITED NATIONS [1958], *Integrated River Basin Development* (New York), 60 p.

UNITED NATIONS [1960], *A Case Study of the Damodar Valley Corporation*; Flood Control Series No. 16 (New York).

UNITED NATIONS [1962], *A Case Study of the Development of the Kitakami River Basin*; Flood Control Series No. 20 (New York).

UNITED NATIONS [1966a], *Compendium of Major International Rivers in the ECAFE Region*; Water Resources Series 29 (New York).

UNITED NATIONS [1966b], *Fourth Biennial Report on Water Resources Development*; 40th Session, Supplement No. 3 (New York).

UNITED NATIONS [1967], *Statistical Year Book 1966* (New York).

U.N.E.S.C.O. [1951a], *Water Resources*; Vol. 4 (New York).

U.N.E.S.C.O. [1951b], Conservation and utilization of resources; Vol. 3 of *Fuel and Energy Resources* (New York).

U.S. DEPARTMENT OF AGRICULTURE [1962], *Land and Water Resources* (Washington, D.C.).

U.S. GOVERNMENT [1950], *A Water Policy for the American People;* 3 vols. (Washington, D.C.).

U.S. BUREAU OF RECLAMATION [1941], *Columbia Basin Joint Investigations* (Washington, D.C.).

VENDROV, S. L. *et al.* [1964], The problem of transformation and utilization of the water resources of the Volga River; *Soviet Geography*, 4, 23–34.

WHITE, G. F. *et al.* [1962], *Economic and Social Aspects of the Lower Mekong Development*.

WHITE, G. F. [1963], Contributions of geographical analysis to river basin development; *Geographical Journal*, 129, 421–32.

World Almanac [1967] (New York).

YOUNG, L. L. [1955], Developed and potential water power in the world; *U.S. Geological Survey Circular* 367 (Washington, D.C.).

11.III. Long-term Trends in Water Use

MARTIN SIMONS

Department of Education, University of Adelaide

1. The growing need for water

It has been reliably estimated that consumption of water in the United States alone will have risen before 1980 to an average of well over 400 billion (400,000,000,000) U.S. gallons per day, an increase of more than 30% over the present requirement, and there is every prospect of continued massive escalation of demand beyond that time. The total fresh water available in the United States from rain, rivers and ground water, even supposing all could be fully used, is about 3.5×10^{14} U.S. gallons per year. At the present rate of increase the United States will have outstripped available resources long before the middle of next century. The rising demand comes from all quarters (fig. 11.III.1). Domestic consumption is increasing, but is not the major problem. More water each year is used for irrigation, not only in arid regions but in more humid areas, where yields can be increased and made more reliable by supplemental irrigation. However, industrial uses of water are now becoming dominant, and by 1980 it is expected that more than half the water used in the United States will go to power stations and factories. Further than this, by the end of this century it is predicted that the most important single user of water will be the thermal electricity generating industry. This water is mostly used for cooling steam condensers for which fresh water is preferred and, in any case, central locations are superior to coastal ones to cut down electricity transmission costs. Nor does the increasing expansion of nuclear-power generation decrease the general water need, since up to the present time such power stations convert the energy of the atomic pile to steam, which is then used to drive the dynamos. In manufacturing industries water is employed in many ways, as a cleansing agent, as a constituent of some products, such as beverages, soaps and detergents, dyes, and various other chemical products, but cooling is still the major use. Already water shortage is hindering development of established industries in many places, and in these same areas the industrial pollution of rivers and lakes is creating its own particular problems.

The situation now facing Western Europe and the United States foreshadows developments to be expected in all industrial regions in the foreseeable future. Already in some parts of Western Europe the supplying of great conurbations like Manchester, Liverpool, London, and the Ruhr has proved much more troublesome than expected. Reservoirs constructed in Wales, the Lake District,

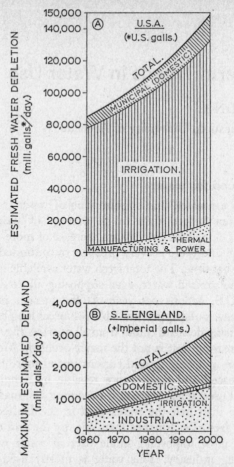

Fig. 11.III.1 Estimated fresh-water depletion for the United States (A), and maximum estimated demand for fresh water in South-East England (B) up to A.D. 2000. These totals do not include water used for hydroelectric generation which is immediately returned to surface storage. It should be emphasized that all values are only estimates, which may be greatly in error, particularly for the more remote dates.

and the Pennines have aroused strenuous opposition, and the cost of water, in cash and in loss of other amenities, is growing with the demand.

2. Geographical engineering

Over and above the accelerating expansion of water-using industries in the populated lands, the great deserts and semi-arid portions of the earth might sooner or later be developed, and water is unquestionably the first essential for this. These are continent-sized areas; if water can be made available they will hold continental populations and provide continental markets. This means water for new industries and new cities as well as for food production. To meet these needs, great new engineering works are required that will make the largest

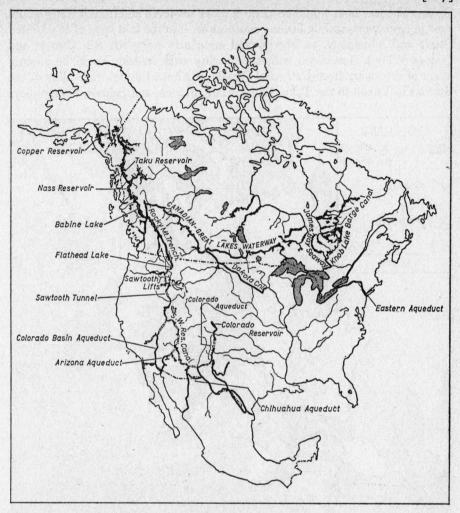

Fig. 11.III.2 The North American Water and Power Alliance proposals (Courtesy of the Ralph M. Parsons Company, Los Angeles and New York).

existing water-control schemes look puny. In retrospect it may turn out that projects like the NAWAPA scheme will be judged too small, rather than too ambitious.

The NAWAPA scheme has been described as 'plumbing on a continental scale', and envisages the control of all the major western rivers of the North American continent from Yukon to northern Mexico (fig. 11.III.2). The centre-piece of the scheme, upon which the rest largely depends, would be the creation of a 500-mile-long reservoir in the Rocky Mountain Trench from Flat-head Lake in Montana across the Canadia border into British Columbia. This would be a result of damming the upper reaches of the Columbia, Kootenay, and Fraser

Rivers. The lake floor would be 3,000 ft above sea-level, and from it water would flow in aqueducts, mainly leading southwards, into the arid west of the United States and, ultimately, to Mexico and aqueducts along the Rio Grande and Sonora Valleys. Associated with this mighty artificial lake would be another chain of reservoirs from Cathedral Rapids in Alaska through the valley of the Stewart in Yukon to the Taku, Liard, Mass, Skeena, and Salmon River valleys

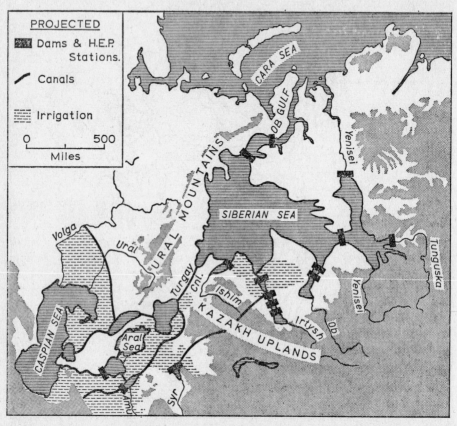

Fig. 11.III.3 One of the plans proposed for the irrigation of Central Asia with the waters of Siberian rivers.

in British Columbia. From these, some water would be pumped to the Peace River Reservoir, and this would feed a great navigable canal bringing water to the Canadian Prairies and, via a canal across the border, to the Dakotas and the upper Mississippi, while a Great Lakes Waterway would connect Lakes Winnipeg and Manitoba with Lake Superior, maintaining a constant water level in the Great Lakes as far as Buffalo, with an aqueduct to New York following the line of the Erie Canal. Associated projects in Labrador, Quebec, and Ontario, and the drawing into the scheme of many smaller local projects already in existence or under consideration are imagined. Compared with the large Californian Water Transfer Scheme, parts of which have been functioning successfully for many years and other parts of which are still debated, this plan seems colossal. There would be scores of dams, pumping houses, power stations, aqueducts, and navigable canals. The cost at 1966 prices was estimated by a sub-committee of the U.S. Senate at about $800 million; probably an under-estimate. Thirty years, from the date of starting construction, would pass before the project's completion. Yet R. L. Nace pointed out in 1966 that even when finished only 13% of the potential water 'crop' of North America would be under the scheme's control. This, in the year 2000, may be too little and too late.

Comparable in scale is the proposed and long-debated Ob-Yenisei–Irtysh diversion scheme, the effect of which would be to bring water from the north-ward-flowing rivers of western Siberia to supply the Caspian Desert by creating an artificial sea in Asia with a surface area larger than England (fig. 11.III.3). Just as North American geologists have queried the ability of the Rocky Mountain Trench floor to withstand the great additional weight of water above, so Soviet climatologists have made gloomy prophesies concerning the probable deterioration of climate in the steppes south of the proposed flooded regions. A great reservoir on the Lower Ob, while doubtless aiding the arid lands and incidentally providing hydroelectric power for industries, would raise the West Siberian water-table, creating widespread swamps even greater in extent than those that would be drowned. Air masses forming above the lake would be cooled, and crops in the south would suffer. The more spectacular Ob-Yenisei plans have therefore been dropped, but others of equal importance have re-placed them, and the debate continues.

Outside the U.S.S.R. and United States, nothing comparable in scale is under serious consideration at the present time, but schemes already in being or under construction, like the Australian Snowy Mountain Project and the Egyptian Aswan High Dam, are more ambitious than anything imagined in such areas fifty years ago, and it would be strange if still larger plans were not made during the next few decades. The Sahara Desert, for example, might one day be irri-gated, although not as crudely envisaged by Sergel's scheme involving the flooding of about 10% of Africa (fig. 11.III.4).

Some of the above schemes seem like the wilder dreams of science-fiction, and yet they involve no important technical innovations. The engineering would be along well-proved lines, the dams would be larger and the aque-ducts longer, but nothing new in principle would be involved. To some extent the same is true of the many schemes suggested for control and use of the sea. The Dutch are already admired for their ambitious land-reclamation pro-jects, the entirely artificial polders which are being created in what used to be a tidal lagoon, the Zuyder Zee, and the Rhine Delta Scheme, which is currently under construction. It is sometimes forgotten by writers on these works that the Yssel Meer, the remnant of the Zuyder Zee, is now a freshwater lake behind the North Sea Dyke, and hence an invaluable source of water for the growing cities of the Netherlands. In the U.S.S.R. a project larger than this has been suggested for the Sea of Azov, at present more saline than the Black Sea, but capable of

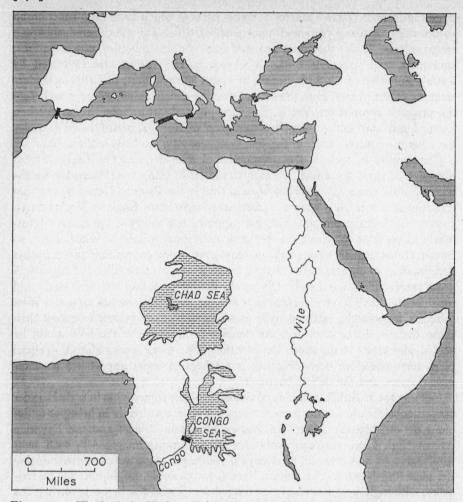

Fig. 11.III.4 The basis for H. Sergel's plan for the irrigation of the Sahara.

being enclosed by a relatively short dyke across the Kerch Strait. There are many other saltwater lagoons where similar projects could be suggested. On the smaller scale, in Britain the Wash and the Solway Firth might all physically lend themselves to such development, providing that the schemes were economically viable. In the United States similar freshwater reservoirs might be created in such inlets as Chesapeake and Delaware Bays, Long Island Sound, San Francisco Bay, and, by damming the Straits of Georgia and Juan de Fuca in Canada, a freshwater lake of great size could be made to supply the industrial northwest. Such plans, of course, cannot be considered in isolation. The enclosure of many of the world's major estuaries would have profound effects upon the cities and other communities around them which depended for their existence on access to the sea. Fishing villages in the Netherlands were forced to change their ways drastically when the Zuyder Zee was enclosed. A major port would insist on retaining access to the oceans, and might gain little from the impounding of its estuarine waters.

On the larger scale, even the enclosure of such vast bodies of water as the Mediterranean Sea, by damming the Straits of Gibraltar, has been proposed, though often, it seems, these ideas are deliberate pipe dreams or academic exercises rather than serious propositions. Nevertheless, under the sharp pressure of demand, even these notions will be taken seriously and will, at least, advance to the point of feasibility studies, if not further. The Mediterranean is somewhat more saline than the Atlantic, and shutting off its western end would if anything increase, rather than reduce, its salinity, which is caused by evaporation from its surface. It might even be expected that the level of the sea would fall and the lake become, like the Caspian, a shrinking reservoir of salt water. Effects on climate might also be considerable, for not only would conditions on the Mediterranean coasts be changed, possibly for the worse, but the cessation of currents through the straits might affect the larger pattern of water circulation in the Atlantic. Similar objections could be raised to some American projects, now abandoned, to divert the Gulf Stream by erecting a barrier between Florida and Cuba, with the intention of increasing the temperature of the Gulf of Mexico and thus preventing the damaging frosts that affect crops in the southern states. Another idea, to divert the cold Labrador current away from the American coast, by building a barrage into the Atlantic south-east of Newfoundland, would probably bring the Gulf Stream closer to American shores, but its likely corollary would be disaster for Europe, which would suffer a drastic fall of temperature. In the same dangerously speculative category must be placed the Soviet scheme to place a barrier across the Bering Strait and then, using a series of giant sluice gates, opened and closed alternately, allowing the flood tide to pass through but to restrain the ebb. In this way it would be possible technically to abstract large volumes of warmer water from the Pacific and pass them into the Arctic Ocean, where they might melt, or help to melt, the sea ice. Alternatively, by subtracting cold water from the Arctic, the Gulf Stream entering from the other side of the Ocean would be drawn farther into the basin, so having a similar effect. The climatic effects of any such scheme, put into operation, are at present quite unforeseeable.

3. Water purification

For any large-scale scheme innovation is necessary in the political and economic spheres, rather than in engineering. The Canadians are by no means so enthusiastic about NAWAPA as the Americans, for some see the project as robbing Canada of her water. Although at present she is not using more than a fraction of the available supply, there will certainly be a time when Canada's own development is restricted by lack of water. In Siberia, too, there are sectional interests opposed to the sacrifice of one region, by flooding or deprivation of water, to the convenience of another, though strong centralized government tends to conceal such divisions of opinion. In Africa political issues already loom very large in water control, and obviously no ambitious scheme for the Sahara can make headway unless the participating nations can agree. Political stability and trust are a primary requirement before any kind of financial arrangements can be made, and it is equally clear that however such enormous engineering works are to be capitalized, something new in the way of long-term investment and re-payment agreements will be necessary before it is worth any draughtsman's time to put pencil to paper. Very probably it will be difficulties in these realms, rather than in the technical field, that will compel men to turn to other sources of water, and to more careful control of existing supplies. There are two obvious lines that development will take. Water, unlike many other raw materials, is rarely destroyed. It is used and re-used by man at different stages of its journey seawards. However, it may enter the city or factory in a natural state and leave it more or less polluted. Rivers thus carry vast quantities of waste, and even the sea is polluted by sewage, by oil from the bilges of tankers, and by radioactive waste. Man could increase the amount of available usable water by large-scale purifica-tion and recirculation, just as in the closed environment of a space capsule a limited quantity of water is used again and again, purified, recirculated, re-purified, and so on for an indefinite period. A water-using industrial complex on earth could treat its water in exactly the same way, a necessary quantity being abstracted from the natural water cycle in the first place, and thereafter perpetu-ally recycled. Only solid waste would ever leave the site, and little further water would be needed unless some planned expansion of the industry were under-taken, when new water-treating plant would necessarily be included at the plan-ning stage. Although at the present time such closed water systems are not necessary, as a rule, they do exist on a small scale, particularly where water of greater purity than that found naturally is required. One example of an existing partial recycling system is at Bedford, England, where polluted water is taken from the Great Ouse, passed through a complex sequence of processes involving chemical precipitation, flocculation and softening, filtration, chlorination, and so on, before redistribution. Such schemes are expensive, but as demand rises, costs of supplying clean water from natural sources will increase and the political difficulties will loom ever larger, whereas further scientific advances and engineer-ing improvements can be expected to reduce the relative cost of recycling.

Similar arguments will prevail in future with regard to desalinization of sea-water and other mineralized sources. The cost per gallon of water obtained in this way, allowing for changes in the real value of money, is falling as improve-ments are made to equipment and as new sources of power are discovered. It is very probable that within a few decades it will become cheaper, as well as politically preferable, for cities like Los Angeles and London to augment their water supplies by large-scale distillation or other desalinization plant, powered by nuclear energy, or conceivably by cheap solid fuels. A large desalinization plant could be projected, designed, and constructed in a few years at most, and once built would remain under the control of the community which made the capital investment. Neither advantage accrues to the enormous water-transfer projects. Such engineering achievements as the Aswan High Dam and the Californian

Water Plan and the control of the Jordan have been handicapped by political troubles, and this must be a factor in the calculations of governments and investors as water shortage increases. A series of distillation or electrodialytic plants, while seeming less efficient overall, might be considered a more practicable investment, if locally demanded and locally financed, rather than schemes involving withdrawing resources from regions where their loss might later be deplored.

4. Conclusion

The past provides many examples of forecasts and projected trends which have proved wrong, but so far there are no signs of any tailing off of demand for water, and no suggestion of any important substitute being found. On the contrary, new industrial processes and new inventions frequently appear which increase the reliance of civilization upon this fundamental resource. Some of the expedients that will be necessary to meet the challenge will doubtless be extraordinary, others will be merely extensions and expansions of existing engineering methods. In the long run water supply may prove to be the ultimate limiting factor for world population. Synthetic food is a possibility, even a probability for the future feeding of the masses, but manufacture of new water by chemical means, though technically possible, can hardly be contemplated on a scale sufficient to make any substantial difference to the main problem. The total amount of water available to the race is virtually fixed. In the long term it is the rate at which the liquid can be passed through the biosystem that will set a limit to the development of human society. If all known sources of fresh water were fully used about 4.5×10^{15} (Imp.) gallons would be available, or 20,000 k^3. This would imply a world population of 20,000 millions at roughly the present *per capita* rate of consumption. Some estimates suggest that by A.D. 2100 world population will have reached or passed this figure if the food problem can be solved. After that, water must come from the sea.

REFERENCES

ACKERMAN, E. A. and LÖF, G. O. G. [1959], *Technology in American Water Development* (Resources for the Future, Inc., Baltimore), 710 p.

ADABASHEV, I. [1966], *Global Engineering* (Progress Publishers, Moscow), 237 p.

ANON [1960], New water treatment works of the Borough of Bedford Water Undertaking; *Water and Water Engineering*, **64**, 293–9.

CAMPBELL, D. [1968], *Drought; Causes, Effects, Solutions* (Cheshire, Melbourne).

DIESENDORF, W., Editor [1961], *The Snowy Mountains Scheme: Phase I* (Horwitz Publications, Sydney).

FURON, R. [1967], *The Problem of Water; A World Study;* Translated by P. Barnes (Faber and Faber, London), 208 p.

INSTITUTION OF WATER ENGINEERS [1967], Symposium on conservation and use of water resources in the United Kingdom; *Journal of the Institution of Water Engineers*, **21**, 203–330.

INTERNATIONAL UNION OF PURE AND APPLIED CHEMISTRY [1963], *Re-Use of*

Water in Industry; Committee of the Water, Sewage and Industrial Wastes Division, (Butterworth's, London), 256 p.

MILLER, D. G. [1962], *Desalination for Water Supply* (Water Research Association, Medmenham, Bucks).

NACE, R. L. [1966], Perspectives in water plans and projects; *Bulletin of the American Meteorological Society,* **47,** 850–6.

SYMPOSIUM [1964], *Water Resources Use and Management;* Australian Academy of Science (Melbourne University Press).

THE RALPH M. PARSONS CO. [1967], *North American Water and Power Alliance* (Los Angeles and New York), Brochure 606–2934–19.

12.I. Choice in Water Use

T. O'RIORDAN

Department of Geography, Simon Fraser University

and

ROSEMARY J. MORE

Formerly of the Department of Civil Engineering, Imperial College, London University

1. Traditional allocation

Water has been used by man in many ways. His first need was for a drinking supply for himself and his animals, then navigation on the waterways allowed him to move quickly and cheaply from place to place. In the drier climates water was early used for irrigation, in wetter, low-lying areas later drainage schemes made the land habitable. Man has long been the victim of disastrous floods, which today he is increasingly able to combat by flood-warning devices and reservoir storage. Currently water is also managed for a great variety of industrial purposes, for recreation and for its natural beauty in the landscape, all of which require maintenance of water quality and control of pollution.

The number of uses to which water is put varies in time and space, and the possibility that future new uses of water will be developed means that sufficient water should be left in long-term water inventories to accommodate them. In any specific region one or two water uses are often dominant, whereas others are subsidiary or unimportant. In many areas water uses conflict with one another (e.g. domestic water supply and irrigation), whereas some demands are complementary (e.g. pollution abatement, flood prevention, and recreation). Whether the many uses of water are conflicting or complementary, water management by allocation is necessary.

The classic means of allocating water between users has been by water rights. The oldest of these rights is the *riparian doctrine*, which was part of the civil law of Rome and later became incorporated in the common law of England. In brief, the riparian doctrine accords to the owner of riparian land, by or across which a stream flows, the right to use the water of that stream on or in connection with his contiguous land. It is not based upon use of water, and is not lost solely by disuse. Under English common law riparian rights may be altered by grant or prescription, for example, under the 'reasonable use' doctrine which states that riparian users must give consideration to reasonable demands by others with riparian rights. The *prescriptive right* to take water may be acquired through actual and uninterrupted use, generally for a long period, with acquiescence of the person from whom the right is acquired. All private industrial water supply from rivers in England and Wales, together with much irrigation, depends on prescription.

Whereas the land-based riparian and prescriptive rights were adequate in humid European conditions, they proved unmanageable when European immigrants began to farm and develop industries in the drier parts of the United States and Australia, and an *appropriative doctrine* was developed founded on the principle of temporal priority of diverting water and putting it to beneficial use upon land ('first come, first served'), regardless of the contiguity of the land to the source of supply. Appropriative rights, which are based upon use of water and are lost by non-use, allowed the spread of irrigation away from the river courses of the western United States.

Of more recent years, the allocation of natural water resources has rested largely upon the political process. In the United States, in particular, large-scale development schemes were the outcome of a combination of three factors: crises (usually natural catastrophes, though that preceding the New Deal 1933–9 was an economic one); a dominating political personality; and identification by vested-interest groups with the major federal agencies responsible for the implementation of large projects (dams, hydroelectric schemes, etc.). With the notable exception of the Tennessee Valley Authority, water-resources development during this period tended to be more project-orientated rather than conceived in terms of comprehensive resources management. Since dam construction meant employment, and stored water brought such varied benefits as flood protection, irrigation, and hydroelectric power, few politicians could afford to resist the temptation to attract federal capital into their constituencies, regardless of how uneconomic or non-integrated the project might turn out to be. The practice which developed whereby politicians helped each other to push water-resource projects through Congress was known as 'pork barrel legislation'.

2. The need for modern allocation techniques

As the demand for water grew both in relation to absolute requirements and with regard to the variety of uses to which it could be put, so unmanaged natural water-resources have become less able to supply these needs. This has been particularly the case in the drier western parts of the United States, where demands could be met only by large-scale transfers of water, and, even in the more humid regions, transfers of water resources, both in terms of space and function (e.g. change from irrigation to municipal use), are occurring, necessitating the introduction of rigorous techniques whereby water can be allocated both in space and in time in an optimal manner. Two major concepts in water-resources management have helped to emphasize the need for improved allocation techniques: those of multi-purpose use and of integrated river-basin development.

Multi-purpose use involves the simultaneous management of water resources (river basin, ground water, lake, reservoir, etc.) to produce a variety of functions, such as flood protection, public water supply, irrigation, hydroelectric power, navigation, pollution abatement, outdoor recreation, fishing and wildlife preservation, and the like. Each of these functions is spatially linked to a complementary use of land and economic activity. For example, water is consumed by activities on the land (e.g. irrigation) and urban complexes discharge waste products into the rivers, both of which are necessary to maintain the local economy. Therefore, the problem of allocating multi-purpose water resources also includes the important spatial element of land-use planning. *Integrated river-basin development* can be viewed in terms either of single-purpose development (e.g. flood protection, as in the Miami Conservancy in Ohio) or multi-purpose development, but in both cases the natural-resources base encompasses the total hydrological unit of the river basin. This concept again introduces very important spatial and temporal aspects, for any development of water resources in the headwaters of a river must undoubtedly affect the quantity and timing of downstream supply.

Present policy in relation to the need for large-scale spatial transfers is tending towards the concept of *multiple purpose re-use*. A 'use' or 'purpose' can be defined as the combination of specific productive inputs and the resulting outputs of goods and services. Traditionally, a distinction has been made between 'consumptive' uses, where water was actually used up (e.g. irrigation, public water supply), and 'non-consumptive' uses, where a certain volume of water was required but little or none was actually lost in the performance of the function (e.g. recreation, pollution abatement). In strict economic terms all uses are consumptive to a degree, whereas in physical terms most uses are largely non-consumptive, and a more accurate distinction can be made between those functions which involve a direct transfer of water outside the channels (e.g. public water supply and irrigation), which might be termed 'transferred' functions, and those which provide benefits through their presence in the channel (e.g. navigation, pollution abatement), which might be designated *in situ* functions. Multiple purpose re-use involves the spatial transfer of additional water supplies (e.g. ground water or surface water stored behind a dam) by means of the natural river channel. Not only is it much cheaper to utilize the natural channel as an aqueduct but by maximizing the use of water over space considerable economies may be achieved. Such a policy permits greater development of *in situ* functions, stimulates economic development throughout the river basin and its spread from one basin to another, yet still satisfies the demand made by 'transferred' functions. The 're-use' concept relates to the water abstraction by 'transferred-use' consumers from the river channel and its subsequent return after suitable treatment to the natural channel, though possibly at a different location (e.g. purified sewage effluent from the municipality).

The basin-water-resource base can be envisaged therefore as a system of interlinked multi-purpose demands (fig. 12.1.1), some conflicting and others complementary, which is managed by means of projects. A 'project' may be defined as a given set of inputs designed to produce the required combination of outputs from a resource system. The purpose of allocation techniques is, therefore, first to discover the optimum mix of functions for any given project in order to maximize net benefits (e.g. gross benefits minus gross costs), and secondly, to weigh-up the relative merits of alternative projects designed to achieve a similar set of productive outputs. In view of increasing demands upon both 'transferred' and *in situ* functions, and the effects upon both spatial linkages and economic activity, the need for a rigorous analysis of choice in water use is becoming more and more important.

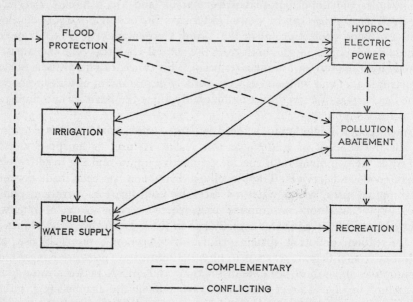

Fig. 12.1.1 The inter-linked multi-purpose demands from the basin water-resource base.

3. Economic and spatial problems in allocation techniques

In an interrelated multi-purpose water-resources system the evaluation of its component functions in precise economic terms is rendered more difficult on account of the highly intricate economic relationships and the spatial linkages involved. The latter are particularly concerned with 'collective benefits', 'externalities', and 'scale', whereas those of primarily economic interest are 'intangibles', 'opportunity costs', 'interest rate', and 'uncertainty and risk'.

Collective benefits

Certain benefits accruing from resources management are universal in character in the sense that they must be provided for a group regardless of individual preference or individual evaluation of those benefits (e.g. flood protection, pollution abatement, fishery protection, wildlife preservation, and the enhancement of aesthetic attractiveness). Since individuals cannot state their preferences or valuation in the market-place through the usual procedure of willingness-to-pay, collective benefits can be distributed and paid for only on a unified areal basis. One method of evaluating collective benefits on an economic basis is to calculate the least-cost alternative to provide the same supply.

Externalities

Water-resource uses are too closely interlinked to permit separate management of any one sector without impinging upon the production and development of at least one of the other sectors within the total complex. Such an interrelationship is called an externality, being positive where the relationship is complementary (e.g. the release of protectively-stored floodwater may dilute downstream pollution and be harnessed to generate hydroelectricity), and negative where the management of one function impinges upon the benefit outputs of another (e.g. the loss of aesthetic satisfaction in a recreation area due to the excessive discharge of polluting effluents). Externalities usually have a strong spatial component, for the effects on one function may be spatially removed from the management of the other. Thus a reservoir provides benefits both on site (recreation, power, water supply) and downstream (flood protection and pollution dilution).

Scale

The identification and calculation of benefits accruing from a project depend to a large extent upon its geographical and economic scale, and a project large enough to bring about repercussions throughout the entire economy must be treated differently from a regional resource development project, where secondary effects, particularly the extent to which they affect the local economy (i.e. the 'multiplier effect'), assume a greater meaning.

Intangibles

In any economic allocation process some common yardstick of evaluation is necessary by which to compare production inputs and outputs. The usual criterion is the tangible price of the goods in a free market. However, some goods (called 'intangibles') are essentially subjective or difficult to identify in monetary terms, such as the visual beauty of a landscape, the protection and preservation of a unique ecological habitat, or the maintenance of a clean river. Owing to the inherent difficulties of evaluation, intangibles are usually underemphasized or overemphasized, depending upon the personal attitudes of the decision-makers.

Opportunity Costs

Opportunity costs are of two kinds: the value of benefits forgone to one resource sector of the project owing to the selection of an incompatible alternative proposal (e.g. the loss of productive farmland when the headwaters of a river are dammed for the purpose of floodwater retention), and when a limited budget causes the diversion of capital away from other investment.

Interest Rate

The interest rate may be regarded as the opportunity cost of borrowed money on the assumption that increased capital investment in the present will result in a better return in the future. The interest rate at which money is borrowed to finance water-resource management projects is critical both in determining the economic justification of the project and in deciding upon the time distribution of expenditure. From fig. 12.1.2 it can be seen that at an interest rate of 7%, 90% of the benefits accruing from the project will be found within 34 years, while at

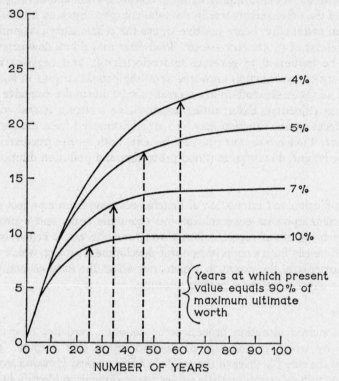

Fig. 12.1.2 Incomes and expenditures of dollars per year present value at various interest rates plotted against number of years. The vertical axis shows present annual value of costs and benefits (in dollars). (From Sewell et al., 1961.)

an interest rate of 4%, 90% of the project benefits will only be realized after 60 years. In the private sector of the economy, where the time horizon is short, interest rates are high, but in the public sector (where most resource-management projects are financed) a greater degree of attention is paid to benefits accruing to future generations, and interest rates are lower. In the United States the current interest rate for financing resource-management projects is approximately 3%.

Uncertainty and Risk

Both uncertainty and risk also reflected by the interest rate. Increasing risk, the gamble that is taken in the face of unknown and unpredictable supply and demand, usually results in the raising of the interest rate. Uncertainty involves such possibilities as the discovery or invention of new processes which might render an existing process uneconomic. For example, desalinization, coupled with nuclear-electric-power production, might seriously impair the long-term feasibility of projects designed to provide hydroelectric power and the long-distance transfer of fresh water.

For the reasons given above, multi-purpose resource-development projects are usually found in the public sector of the economy, it being generally the case that private interests do not take due consideration of all social (i.e. community welfare) costs, since they cannot recoup uncompensated benefits, and market incentives do not function ideally. However, it should not be thought necessarily that the public decision-making process is in any way more perfect than in the private sphere; and recent work in water-resource allocation techniques is concentrating on the role of the decision-maker and his attitudes to, and perception of, the problems that confront him. The decision to develop resources from a public or private standpoint is therefore by no means clear-cut, and the history of resources development has been marked by constant conflict between vested-interest groups whenever a proposal by private interests affects what is generally considered to be 'the public interest'. One of the most notable examples of this in American history was the decision to construct the Hetch Hetchy public-water-supply reservoir along with the production of hydroelectric power in the Yosemite National Park of California. This was treated as a test case against private development in public territory, but after more than a decade of struggle (1901–13) the dam was ultimately constructed, since there appeared to be no cheaper alternative that would provide the same benefits. A similar problem arose more recently in England when Manchester Corporation, facing a water shortage of 50 million gallons per day by 1970, unsuccessfully petitioned for an abstraction from two famous tourist-amenity lakes, Ullswater and Windermere. Such decisions on conflicting issues abound in resource-management history, and it is therefore not surprising that recent workers have become interested in such matters as basis of opinion, value judgement, policy, and perception, which represent the final phase of the allocation process.

4. Allocation techniques

There are two main technical approaches to the problem of water-resource allocation, benefit–cost analysis and systems analysis, the latter being in some respects a more sophisticated and dynamic development of the former.

A. Benefit–cost analysis

Benefit–cost (b–c) analysis is a technique for enumerating, evaluating, and comparing the benefits and costs that stem from the utilization of a productive resource base (a project). The economic theory upon which b–c analysis is based assumes that a desired objective is achieved through the lowest-cost alternative of a number of alternative projects designed to produce a similar objective or, in the case where the quantity of resource use is specified, the most

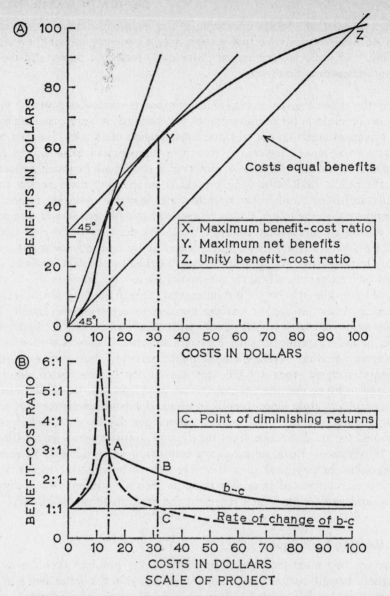

Fig. 12.1.3 Benefit–cost relationships for various scales of development (From Sewell, W. R. D., Davis, J., Scott, A. D. and Ross, D. W., Guide to benefit–cost analysis, In Burton and Kates, 1965, pp. 544–57).

efficient solution of a number of alternatives is that one which maximizes the desired objective. In terms of welfare, b–c analysis does attempt to maximize the net social benefits to a point of project implementation when the benefit accruing to any one set of consumers cannot be further increased without incurring losses upon another set of consumers.

According to b–c theory, the ideal project is the one where total net benefits (i.e. total benefits minus total costs) are maximized (i.e. fig. 12.1.3(a), point Y). At the point X on figure 12.1.3(a), the ratio between benefits and costs is at its greatest for the project, but the scale of the project is not maximized, since an increase of costs (even though costs are now rising in relation to benefits) will still produce sufficient benefits to increase the total net benefits. The optimal scale of development is reached when each component resource sector of the project is developed to the point of diminishing returns, i.e. where the last increment of cost invested just equals the value of benefit it produces (fig. 12.1.3(b), point C). Theoretically, the construction of such a project makes the best use of resources and capital in the sense that no greater net gain could be produced elsewhere in the economy.

The considerable complexity of economic and political issues that beset optimal allocation processes in the real world exert severe constraints upon the theoretical bases of b–c analysis which assume perfect knowledge of the market (i.e. no uncertainty), quantifiable benefits and costs as valued in the market, and an explicit statement of individual preferences. Furthermore, the objective of maximizing net social welfare relies upon equating individual preferences (utilities) and the total aggregation of inter-personal utilities (e.g. individuals' attitudes to water-based recreation). A weakness of conventional b–c analysis is that multi-purpose resource-development projects can be implemented for a variety of objectives, such as redistribution of national income, regional economic growth, and increased aesthetic satisfaction, for which the overriding principles of economic efficiency do not rule. When one asks, 'Whose welfare is being maximized?' and 'Who is paying for this?' it is found that the biggest beneficiaries usually exert the greatest vested interest in the decision-making process, such that these pertinent questions cannot properly be answered in a rigorous economic framework. B–c analysis, therefore, operates in the socio-political arena, where sub-optimum economic decisions must be partly based upon compromise and value judgements. Another difficulty is that b–c analysis often emphasizes the static view of the project, whereas objectives change both with the passage of time and as more data is collected – in other words, there is often considerable feedback (interaction) between policy making and programme design. Also, from a mechanical viewpoint, b–c analysis is such a laborious exercise that only a limited number of alternatives can be scrutinized with the necessary care, so that step-by-step analysis through time is virtually impossible.

Despite these considerable limitations, b–c analysis does have merits as an allocation technique and is still widely used in evaluating the feasibility of water-resource-management projects, because:

1. It is simple to understand and relatively easy to calculate in comparison with other resource-allocation techniques. Also, it does attempt to take the long view in the sense of analysing both short- and long-term benefits and costs, and the wide view by introducing the concepts of social (or collective) costs and benefits, and externalities.
2. By sifting out uneconomic proposals, it does provide a yardstick for measuring the *relative* feasibility of alternative resource-management programmes, whereby public money will achieve the most satisfying return.
3. Assuming that a common procedure for calculation is adopted, b–c analysis does help to establish priorities among a group of alternative resource-development proposals of a similar scale, and may also help in the sequence and timing of projects (e.g. a group of dams in a large river basin).
4. B–c analysis demands the detailed study and evaluation of each of the component resource sectors that constitute a project, and assists in identifying the location, intensity, and frequency of all benefits and costs, thereby helping to pin-point specific beneficiaries.
5. While the evaluation of intangibles is difficult to reconcile within the b–c framework, such costs and benefits are placed in juxtaposition with more tangible ones thereby providing at least some means for their inclusion in the decision-making process.

To sum up, b–c analysis is most relevant in a choice between resource-management projects within one context (e.g. various proposals to develop one river basin or group of basins), where the costs and benefits are relatively easily recognized and calculated (e.g. water-resource management in preference to education or health programmes), and are at a comprehensible scale (e.g. river basin, rather than on the scale of national resources).

The following Tables (12.1.1–3) give a simple example of benefit–cost analysis for a river basin which presents four possible dam sites for multi-purpose development (Spargo, 1961).

Assuming full employment, an interest rate of 5%, a project life of 50 years with no salvage value, and current price levels for all costs, which dam, or combination of dams, will achieve the maximum net benefits?

Obviously Plans 4, 6, and 8 can be quickly excluded, since their b–c ratios

TABLE 12.1.1

	Construction costs ($000)	Operation and maintenance ($000)
Dam A	10,000	30
Dam B	5,000	20
Dam C	3,000	20
Dam D	4,000	20

TABLE 12.1.2

	Benefits ($000)	Plan number
Dam A alone	1,000	1
,, B alone	400	2
,, C alone	200	3
,, D alone	100	4
Dams A + C	1,300	5
,, B + C	400	6
,, D + C	500	7
,, B + D	500	8
,, B + D + C	900	9

TABLE 12.1.3

Plan number	Construction costs ($000)	Present worth annual costs ($000)	Present worth total costs ($000)	Present worth total benefits ($000)	b–c ratio
1	10,000	549	10,549	18,300*	1·74
2	5,000	366	5,366	7,320	1·35
3	3,000	366	3,366	3,660	1·09
4	4,000	366	4,366	1,830	0·41
5	13,000	915	13,915	23,790	1·71
6	8,000	732	8,732	7,320	0·84
7	7,000	732	7,732	9,150	1·19
8	9,000	732	9,732	9,150	0·94
9	12,000	1,098	13,098	16,470	1·27

* The factor 18·3 used to generate the total benefits is derived from the 5% interest rate over a fifty-year time span (Kuiper, 1965, p. 408).

fail to reach unity. Although Plan 1 appears to have the most favourable b–c ratio, its scale of development will only be optimal if its *marginal* benefits equal *marginal* costs (fig. 12.1.3(b), point C). The method of testing for this is to compare the benefits and costs of the project with the lowest total costs with that having the next lowest total costs (because economic efficiency aims at minimizing total costs). Thus, for example, comparing Plan 2 with Plan 3:

$$\frac{\text{Benefits of Plan 2} - \text{Benefits of Plan 3}}{\text{Costs of Plan 2} - \text{Costs of Plan 3}}$$

$$= \frac{7,320,000 - 3,660,000}{5,366,000 - 3,366,000} = \frac{3,660,000}{2,000,000} = 1·83$$

The ratio 1·83 represents the marginal benefit–marginal cost ratio comparing Plan 2 with Plan 3.

Plan 2 is therefore superior to Plan 3, since, despite greater costs, the incremental benefits thereby achieved are higher, and hence net benefits are greater. Table 12.1.4 continues this analysis.

TABLE 12.1.4

Plan comparison	Incremental costs ($000)	Incremental benefits ($000)	Incremental b–c
3 alone	—	—	1·09
2 over 3	2,000	3,660	1·83
7 over 2	2,366	1,830	0·78
1 over 2	5,183	10,980	2·12
9 over 1	2,549	1,830	0·72
5 over 1	3,366	5,490	1·67

Thus, by elimination, Plan 5 (i.e. the joint construction of dams A and C) is superior to Plan 1 (i.e. dam A alone) in terms of maximizing net benefits.

B. Systems analysis

A system is a unit composed of interacting parts, whose qualities and relationships can often be conveniently expressed in mathematical terms (by means of a model) and analysed as a whole. River basins obviously can be viewed as open, self-regulating physical systems maintained by a throughput of water. However, each river basin is hydrologically unique and demands a fundamentally different systems model for its analysis. Within this varied hydrological framework of channel geometry, hydrological events (e.g. storm rainfall) and basin responses (e.g. surface runoff), there is often a great variety of possible dam sites, the crucial building blocks of most water-resource systems, their respective sizes (i.e. a few big or many little ones), whether they operate in series or in parallel, the positions and sizes of associated hydroelectric plants, the maintenance of sufficient reservoir capacity to combat possible floods, the maintenance of sufficient channel flow for irrigation, navigation, pollution control, etc. The major problems in basin planning are not so much where to send the water but how much and at what time to do it.

Where economic development involves the control of a hydrological system a more complex hydro-economic system results, wherein economic and social variables interact with, and constrain, the purely hydrological ones, giving rise to the now-classic geographical notion, exemplified by T.V.A., of the fusion of a hydrological system with a coherent socio-economic system. The description of such a system involves the specification of:

1. The nature of the hydrologic and socio-economic inputs (e.g. rainfall, runoff, population statistics, demands for water from various users, etc.);
2. The nature of the outputs (e.g. water supply, electrical energy, etc.);
3. The variables involved (e.g. reservoir capacity, size of power plants, etc.); and
4. Some mathematical statement (model) relating inputs, outputs, and system states in time. The state of the system is its instantaneous condition, which is characterized by its composition, organization, and energy flows, and

defined by system parameters. Hydro-economic systems are thus highly complex and contain many elements that are difficult to quantify (e.g. constraints on the system arising from political and institutional factors). Nevertheless, the development of electronic computers has enabled quite complex systems of variables and their relationships to be treated.

There are four main steps in the design of a systems analysis for an integrated river-basin development:

1. Specification of objectives.
2. The translation of the objectives into design criteria.
3. Field-level planning.
4. Comparison of the results with the objectives.

Specification of the objectives

As has been seen, economic and social objectives may vary widely and involve such broad considerations as increasing the overall economic and social efficiency of the region, the income *per capita*, and the employment opportunities. Systems analysis differs from benefit–cost analysis, therefore, in that its aims may not be narrowly financial (e.g. the net profit to be made from engineering structures), but in some respects its accounting procedures are similar to, and represent an extension of, b–c analysis. Thus, where possible, the collective objectives are quantified and specified as an *objective function* whose value is to be maximized, a simple example of which can be expressed as follows:

$$\sum_{t=1}^{T} \frac{E_t(y_t) - M_t(x_t)}{(1 + r)^t} - K(x) \tag{1}$$

where $E_t(y_t) =$ gross efficiency benefits in the tth year from the output (y) of the system (x) in that year;

$M_t(x_t) =$ operation, maintenance, and replacement (OMR) costs in the year (t) for the constructed system;

$K(x) =$ capital costs for the constructed system;

$r =$ discount rate;

$T =$ length, in years, of the economic time-horizon.

The summation yields the present value of the continuing stream of gross benefits, less OMR costs, for a period of t years, discounted at an interest rate of r. Subtraction of the undiscounted capital costs (assumed to be incurred in year 1) yields the present value of net efficiency benefits, which is the quantity to be maximized.

The translation of the objectives into design criteria

This involves such processes as mathematically specifying the relative importance of the objectives, the relationships between the various attributes of the completed system, any limiting constraints on these attributes (e.g. maximum physical size of the possible dams, maximum capital available at a given stage of the development, minimum dry-season river flows, etc.), together with discount rates and opportunity costs. The most important relationships are the *cost–input*, *benefit–output*, *production*, and *capital cost–output* functions (fig. 12.1.4).

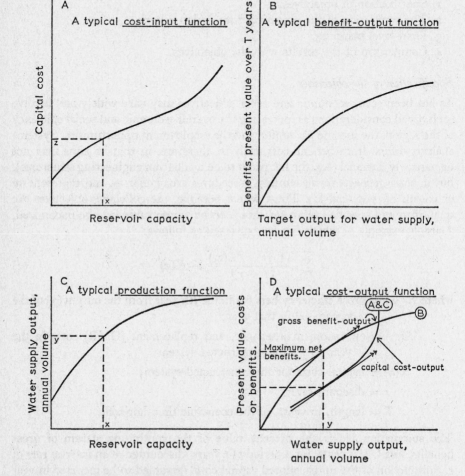

Fig. 12.1.4 The relationships between some of the attributes of a water-resource system (After Hufschmidt, M. M., The methodology of water-resource system design. In Burton and Kates, 1965, pp. 558–70).

A. A typical cost–input function.
B. A typical benefit–output function.
C. A typical production function.
D. A typical cost–output function.

Field-level planning

This aims at maximizing the value of the objective function, subject to the above relationships and constraints. At this stage the computer juggles with all the variables in the objective function, maintaining the relationships and constraints expressed by the design criteria, to produce one or more alternative allocations of water to fulfil the initial objectives. Obviously the large number of variables involved, their range of hydrological and economic character, and the complexity of their superimposed relationships and constraints makes their mutual optimization a highly complex matter, even with the aid of computers. This optimization can be attempted either by constructing mathematical *analytical models* or through *simulations* of the system. These are both *paper experiments* wherein the specimen inputs, storages, states, relationships, and outputs of the system are specified and analysed mathematically. This allows for the simultaneous consideration of such variables as monthly averages and probable variations of streamflow, the capacities of the contemplated reservoirs, the amount of HEP needed month by month, the changing probable needs for irrigation and industrial water, etc.

Analytical models grossly simplify the system down to a few parameters in an attempt to discover optimum values for the most important relationships within the system. Two valuable techniques in this connection are linear and dynamic programming. Linear programming is an optimization procedure to find the maximum or minimum value of a combination of linearly-related variables subject to a number of constraints on the values which they may take. Dynamic programming is a more time-oriented mathematical approach to problems involving an optimum sequence of decisions. It assumes that whatever the initial state and initial decision may be, the remaining integrated decisions constitute an optimum policy with regard to the state resulting from the first decision. The chief value of analytical models in water-resources-system analysis is that they can be used to develop a set of manageable mathematical relationships which can be solved to indicate the range of variation within which a reasonably optimum solution can be obtained by exploration, often by a sequence of simulations.

It is difficult to build the time element satisfactorily into analytical models, although less so than into b–c analysis. In contrast, the *simulation model* can accurately express the step-by-step changes in element states and interactions during each small segment of time of operation by sets of Markov type, or differential, equations. This allows that the states at the end of one time interval can form the input, under differing conditions (e.g. of investment, demand for water, etc.) if necessary, into the start of the next time interval. In such a flexible simulation model the events occur in the same temporal sequence as their counterparts in real life, and in practice each time interval represents a dynamically programmed analytical model providing an output which is the basis for the input for the next time interval, and so on. Thus an initial set of conditions can be investigated in conjunction with a variety of design parameters proposing many alternative relationships and changing conditions through time. The simulation can be run again and again on the computer with different combinations of design parameters, approaching optimization by whittling away at the possible combinations of states, variables, changes, etc. Hufschmidt and Fiering [1966] have shown how a simulation analysis was used in a water-resource-system development plan for the *Lehigh River*, a tributary of the Delaware in Pennsylvania. The design criteria involved the construction of up to six dam sites to provide regulated flows for water supply (domestic and industrial) in the Bethlehem area; recreation at the reservoir sites; dam storage for flood-damage protection, and storage and head for hydroelectric power generation. There were 42 major design variables, 16 dealing with possible physical facility components (i.e. sizes of 6 reservoirs, of 9 power plants, and the construction of a diversion channel), together with 24 variables relating to the allocation of reservoir capacity. In addition there were 12 monthly values of

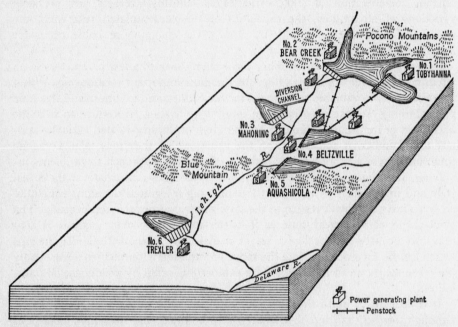

Fig. 12.1.5 The possible physical facility components for the development of the Lehigh basin (Reservoirs 1 and 2 are alternatives) developed by computer simulation. Optimum amounts of reservoir storage, channel flow, hydroelectric output, etc., are calculated by systems analysis (From Hufschmidt and Fiering, 1966).

flood storage and recreation allocation, and 2 output design variables – the target volumes of water supply and electrical energy. The model was programmed so that investments in reservoirs, power plants, and recreational facilities could be progressively scheduled to assumed changing levels of demand. Up to 10 changes in investment and target-output levels could be accommodated, adding 10 time-associated variables, making a total of 52 major design variables in all.

Comparison of the results with the objectives

Even allowing each of the 52 major design variables to have only 3 states ('high', 'middling', and 'small'), this yielded 6 million, billion, billion possible design combinations for the Lehigh River programme, so that a decision was reached after rigorous sampling and testing of the more likely solutions (involving mean and standard deviations of benefits from water supply, energy, recreation, and the prevention of flood losses), together with the examination of each in terms of the initial objectives. Figure 12.1.5 is a sketch of the 16 possible physical facility components.

5. Institutional aspects of water-resource allocation

Allocation techniques suggest how it might be possible to maximize the net social benefits stemming from any given set of inputs, and provide a common economic basis for choosing between various alternative proposals for developing the water-resource complex. However, the decision as to the nature and timing of any particular project rests with what can loosely be described as the political institutional process – that complex of vested-interest group lobbying, public emotion, personality clash, and protested compromise that takes place in committee rooms and council chambers throughout the world. The decision-making process itself is still only imperfectly understood, but it is certainly dependent upon the political institution, for although the allocation of multi-purpose water and land resources involves certain common problems, the final outcome (i.e. implementation of a plan) will differ according to the general policies and attitudes of the nation in question.

Most democratic countries attempt to emphasize the economic significance of the allocation process, although their decisions are often affected by poorly co-ordinated executive powers of governmental agencies and by the pre-existing social and legal institutional arrangements. The presence of these 'non-optimal' factors in the real world introduces practical difficulties which inhibit the full realization of any optimizing allocation technique and the achievement of complete economic efficiency. Water-resources policy, therefore, will not only vary significantly from nation to nation but even between regions because of the differing political, social, economic, and psychological stresses that are placed upon decision-makers. In reality, therefore, decision-makers are faced with a large number of constraints which must be considered in reaching any conclusion regarding water-resources development:

1. *Physical constraints.* Any resource base must have a physical limit of development, regardless of the total economic inputs.
2. *Fiscal constraints.* Total development will depend upon the amount of public money available in the light of other demands.
3. *Policy constraints.* Certain water-resource functions may not be permitted, either for social or political reasons, to drop below a clearly defined quantity of use (e.g. public water-supply requirements and pollution control).

4. *Legal constraints*. Water-resource legislation is usually the outcome of a compromise between the demands of vested-interest groups. An example of the favouring of vested interest would be the financial encouragement of irrigation in the western United States.

5. *Administrative constraints*. The formulation of an optimum water-resources allocation policy depends upon the willing coordination between existing local authorities.

6. *Ownership constraints*. The management of water resources inevitably involves the management of associated land resources, and opposition by private owners may inhibit or prevent project development.

7. *Quantification constraints*. The constraints outlined above assume greater importance in view of the inherent difficulties of quantifying many intangible benefits and costs, particularly of a social or aesthetic character.

8. *Perception constraints*. The behavioural outcome of choice in water use is affected by the perception of the range of choice available to the decision-maker – be he a direct beneficiary, engineer, or resource planner. Perception of alternative uses in water-resource management is still imperfectly understood, but is now considered to be of great importance in the allocation process.

Three national examples serve to illustrate some different institutional approaches to water-resource-allocation decision-making.

Policy-making and the allocation process in the *United States* is divided between the theorists and the pragmatists. The theorists adhere to the economic principle of *marginal productivity*, which states that the cost of providing the last unit of water should be equal to that which the customer is prepared to pay, so that the different users will adjust their various demands, and for all uses marginal productivity will become equal. Theoretically all water would then be used efficiently, the aggregate marginal valuation of the total-water-resource base would be maximized, and the charge for each use would reflect the loss of water to the system and the opportunity costs of its consumption. Such pure economic theory is not implemented in practice. For example, the marginal productivity of water in the western United States is being artificially distorted in favour of irrigation, and it has been estimated that if 10% of the water at present used in large-scale irrigation were to be transferred to municipal and industrial uses, then output in these latter sectors would triple before the end of the century and would constitute a much more efficient allocation of water. However, a change in policy will depend, in part, upon the extent to which the distortion is perceived by decision-makers and more efficient alternatives incorporated. The pragmatists, on the other hand, recognize the importance of ethical and institutional considerations in influencing allocation policy-making towards preserving and/or enhancing the quality of the environment through multi-purpose resource-management programmes. They hold that what is evaluated as economically optimum by a computer may not necessarily be so visualized by the human user, whose attitude towards water-resource use is related to his *perception* of the problem in the light of the information available to him regarding the various alternatives that might produce the same benefits. Increasing attention is being paid, therefore, to the perception of alternative choices and to the attitudes held by decision-makers and the public, together with the considerable influence that these have upon behaviour and action in multi-purpose water-resources management.

Present *British* water-resources policy centres upon the 1963 Water Resources Act, before which water-resource legislation had been fragmentary and ill co-ordinated and had largely dealt with local individual water-resource functions as they became apparent. The 1963 Act recognized the need for a complete water inventory and constituted a system of twenty-nine River Authorities in England and Wales charged with conserving, redistributing, and otherwise augmenting the water resources of their area in all their economic and social aspects. These River Authorities are responsible not only for a detailed account of water supply through hydrometric schemes but also for the evaluation of demand by instituting a comprehensive system of licensing all water abstractors. The licensing scheme, which is related to a system of charges, has superseded the common law (riparian) doctrine of rights. The Water Resources Board is the national co-ordinating body, advisory in capacity, but responsible for liaison, research, and policy-making, particularly in relation to transfers of water between River Authorities. The Act also saw the need for some overall criterion which would relate the varying seasonal and spatial demands placed upon regional water resources, in proposing the concept of a 'minimum acceptable flow' (MAF) for each main river. The MAF is extremely difficult to establish, since its calculation depends upon the licensing system. Its chief advantage is that it can be used flexibly to set seasonal standards for local amenity and river quality in relation to fluctuating demands. The system of charging for water use is at present being developed and will take into account the amount of water used, the extent to which it is removed from the system (with an adjustment for the location, quantity, and quality of returned water), the nature of the source, the season of the year, and the rate of abstraction. The aim is to weight user charges in relation to the effect of his abstraction upon the total water-resources system and not simply upon initial direct removal. For example, irrigators, who return no water to the system, might be subject to a weighting factor of 100%, but water abstracted for industrial cooling processes may only be levied at 1% of the total amount abstracted, since almost all the initial abstraction is returned to the river close to the point of removal. The charging scheme could thus be used as a form of marginal-cost pricing, in that additional demands for water will contribute indirectly to the capital investment required for providing it, with the weighting functions reflecting the opportunity costs in use set by transferred demands (e.g. irrigation, domestic and industrial water supply). The revenue from charges levied on beneficiaries could prove a very positive factor in financing British water-resources development.

In sharp contrast to both the above institutional bases for water-resource allocation is the *Israeli* legislation, the product of a harsh physical, but even more harsh political, environment. Here all water is considered to be the property of the State under the jurisdiction of a Water Commissioner, who can prescribe norms for the quantity, quality, price, conditions of supply, and use of water, which is regulated by licence for specific purposes only. During periods of shortage Rationing Areas are declared in which uses of water are ranked in the priority – domestic and public purposes, industrial and irrigation.

6. Water-resource systems

One has seen how water-resources management can be visualized in terms of an interrelating complex of functions that constitute a system. Water-resources management is, however, only one such system, and the bonds which unite its component functions are complemented by the links between it and other rural and urban resource-management systems, each of which in turn is composed of interconnecting multi-purpose functions. Regional development plans are often designed to co-ordinate such a complex of resource-management projects (fig. 12.1.6). Water-resources development is essentially spatial in concept, for while the initial programme often relates to a specific drainage system which is usually clearly delimited in physical terms, the benefits may relate to wider areal units, which may be regional, national, or even international in character.

A water-resource *system* is thus an integrated complex of interlinked hydrologic and socio-economic variables operating together within a well-defined area, commonly a drainage-basin unit. Such systems are composed of interlinked *sub-systems*, which are simpler on account of the smaller number and greater simplicity of components, their more restricted areal coverage, and their more restricted aims (i.e. water spreading, ground-water recharge, and human use by wells on an alluvial fan). Water-resource systems are more complex, in that they represent larger-scale integrated systems, in which the hydrological and socio-economic variables are interlocked more complexly within a well-defined areal unit of some magnitude. The large drainage basin being developed in an integrated socio-economic manner as a unit forms the ideal water-resource system. However, modern water needs often demand the creation of *super-systems*, in which, for example, a given socio-economic policy is imposed on a variety of drainage-basin units (each perhaps having very different hydrological characteristics), or in which one very large and uniform hydrological unit is being developed under a number of differing socio-economic bases (usually international). In the former the super-system complexity stems from the need to operate a varying hydrological reality towards some integrated socio-economic goal (e.g. the California Water Plan), whereas in the latter the complexity arises from attempts to operate a large hydrological unit under differing institutional bases (e.g. the Mekong River Plan). The future will obviously see a great expansion in scale of these international super-systems.

The Beech River Watershed, one of fourteen tributary basins of the Tennessee Valley Authority, provides an example of a small water-resources system. On the basis of an Inventory Report on resources and future needs, it is proposed to construct seven multi-purpose dams and one dry dam for flood-

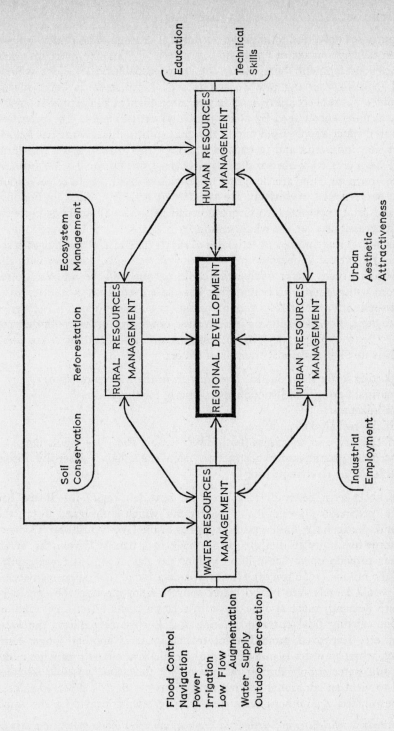

Fig. 12.1.6 The complex of resource-management projects involved in regional water-resource development.

control purposes only, and to improve 80 miles of channel. The resulting flood protection of 17,000 acres of fertile bottomlands will stimulate more intensive agricultural development there, while the steep erodable valley sides will be returned to soil-conserving pasture and forests. More stress is being placed upon a mixed pastoral economy, with larger, more efficient and capital-intensive agricultural units encouraged by controlled rural depopulation. The provision of an assured water supply from one of the multi-purpose reservoirs has helped to attract new industries and to expand existing firms. In all, 2,300 non-farm jobs have been created to absorb the inflow of dispossessed small-scale farmers. The multi-purpose dams are designed to provide a varied range of recreation facilities, some zoned exclusively for specialized activities, such as boating, fishing, and group camping. An improved road network will facilitate convenience and accessibility for the whole region.

In South-East England recent estimates of short- (until 1985) and longer-term (until 2001) domestic, industrial, and irrigation water needs,[1] together with the recognition that many River Authorities could not meet all the future requirements from within their own boundaries, has caused ten River Authority areas to be grouped into a water-resources management super-system. The existence of a future deficiency zone running in a broad sweep from Northampton through London to Ipswich (fig. 12.1.7), has necessitated the consideration of six alternative plans involving regional transfer of water:

1. Importing water from outside (possibly from the Severn or Wye).
2. Estuarinal barrage construction, particularly for the Wash.
3. Desalinization.
4. Artificial recharge of aquifers.
5. Surface storage in reservoirs (both direct supply and river-regulating).
6. Controlled ground-water abstraction from the Chalk (either by direct abstraction or river-regulating).

To meet short-term demands (i.e. over the next ten years) the Board has adopted a 'progressive development programme' which is designed to tap the surface and, particularly, underground supplies of the Great Ouse and Thames. Longer-term feasibility studies are being made of a barrage across the Wash which may provide up to 620 million gallons per day at an estimated capital cost of £300 million (1966 prices). However, a decision of this expensive nature, while it would largely free inland water sources from meeting the considerable future demand, must also be made in relation to its effect on pollution, navigation, existing tidal currents, marine and freshwater ecology, the local economy, etc. Improved ground-water management is another longer-term possibility, whereby water is pumped from the Chalk aquifers to provide additional *in situ* water supply during periods of low flow. One possible scheme, designed to yield an additional 270 million gallons per day, is expected to cost only an estimated £3 million, compared with reservoirs providing the same

[1] Total estimated *deficiencies* in South-East England are 100 million gallons per day in 1971, rising to 400 by 1981, and to 1,100 by 2001.

supply costing between £30 million and £45 million and using up to 10 square miles of land. A technically more sophisticated variant of ground-water abstraction is the artificial recharge of an aquifer whereby water is pumped into the ground-water reservoir during winter periods of excess supply and pumped out again in the spring and summer. However, considerable research is still needed in assessing the hydro-geologic capabilities of the Chalk in southern England

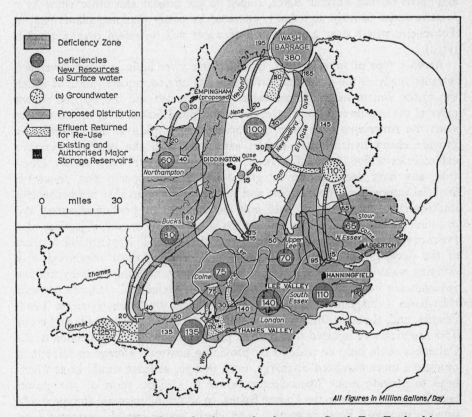

Fig. 12.1.7 One proposed scheme for the supply of water to South-East England in A.D. 2001 involving a Wash barrage, management of the Chalk ground water and surface reservoirs (After Water Resources Board, 1966. Crown Copyright Reserved).

before any major artificial recharge scheme can be put into operation. Of the two remaining proposals, neither inter-regional transfer of water nor desalinization are considered economically feasible in competition with these alternative measures, although technological improvements may render nuclear-powered desalinization, linked to electricity supply, competitive, at least on a small scale. Figure 12.1.7 shows one proposed long-term scheme involving the Wash barrage and increased ground-water use.

An equivalent super-system is being operated under the California Water Plan, which was initiated in 1960 in response to the astronomical rise in State water consumption. The population had increased by 250% in 25 years to 18·5 million in 1965, consuming some 30 million acre-feet of water per year. The Water Plan (fig. 12.1.8) faces the problem that although 75% of the State's water occurs north of San Francisco, 75% of the use is south of it. The basis of the scheme is the construction of a large number of dams, including the Oroville dam (3·5 million acre-feet, generating 644,000 kW of hydro-electricity) and those on the Feather River, linked to the coastal and other cities by a massive series of huge aqueducts, operated by costly pumping plants (e.g. at Tehachapi, where over 4,000 ft³/sec of water will be raised over a 2,000-ft ridge).

Another type of super-system, in which one river basin is being developed internationally, is the Columbia River covering 282,000 square miles, of which the 33,000 square miles of headwaters lie in Canada and the 259,000 square miles of the middle and lower reaches in the United States. Over the past thirty years the Americans have been developing their portion of the Columbia to provide cheap hydroelectric power, associated with the positive economic externalities of flood protection, recreation, irrigation, and water supply. However, any river composes an integrated hydrological system that disregards artificial international boundaries, and it became apparent that comprehensive multi-purpose management could only be achieved by pooling resources and combining proposals for development so as to satisfy mutual objectives at a lower overall cost than the sum of the two national parts. This international aspect of the development raises many problems, including those concerned with differing water laws on either side of the boundary, differing goals by the two governments involved and hence two distinct policies and attitudes towards river-basin management, differing attitudes held by water-resources beneficiaries, and, the most thorny problem of all, the division of benefits and costs. The Americans recognized that the full power potential of their portion of the Columbia could only be realized by providing upstream storage in Canada to produce a more regulated discharge, even though, as there would be too little head to provide much hydroelectric power in Canada, most of the power benefits would accrue to the United States. In addition, upstream storage would provide increased flood protection, and cheap power would encourage local employment downstream, whereas Canada would forgo the real costs of loss of flooded land and the opportunity costs of the future unavailability of the Columbia basin water. It is little wonder therefore that the Columbia River Treaty Protocol Agreement, which was signed by Canada and the United States in 1963, took nineteen years to ratify. The Americans agreed to pay cash to the Canadians as compensation for real costs and to share their downstream benefits of cheap power and increased flood protection. The latter involves payments to Canada of a total of $254·5 million for power benefits and $64·4 million for flood-protection benefits over the first thirty years of the sixty-year treaty (after which Canada can elect to sell her share of U.S. benefits). In return, Canada has agreed to provide 15·5 million acre-feet of storage (including 8·5 million acre-feet for flood-protection purposes) by means of three

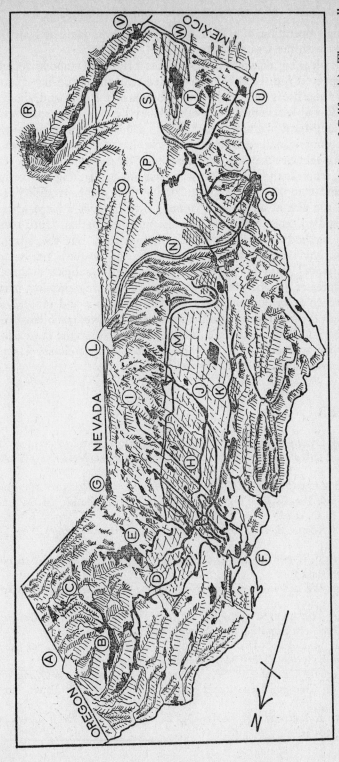

Fig. 12.1.8 California water resources development (from the map issued by the Irrigation Districts Association of California). The small 'boxes' are pumping-stations. (A) Mt. Shasta; (B) Shasta Dam; (C) Mt. Lassen; (D) Sacramento River; (E) Oroville Dam; (F) San Francisco; (G) Lake Tahoe; (H) Hetch Hetchy Aqueduct; (I) Yosemite Valley; (J) San Joaquin River; (K) California Aqueduct; (L) Mt. Whitney; (M) Friant Kern Canal; (N) Los Angeles Aqueducts; (O) Death Valley; (P) Mohave Desert; (Q) Los Angeles; (R) Hoover Dam; (S) Colorado River Aqueduct; (T) Salton Sea; (U) San Diego; (V) Imperial Dam; (W) All American Canal.

projects, although operation of the existing twenty-four dams and all future projects will rest with the United States.

Although the planning of such super-systems may seem impressive, in the long term the greatest future problem associated with the exercising of a conscious choice in water use is the paradox that, whereas on the one hand, significant and economically viable schemes must of necessity be on a large scale and involve complex long-term planning, on the other hand, the increasing tempo of technological, economic, and social changes imply that the vehicles and even the objectives of such schemes may constantly change over comparatively short time intervals. Thus, whereas the vehicles of water-resources planning are becoming more massive and complex, the requirement for their manoeuvrability is also increasing, and the aim of all future planning is to produce a large-scale and completely integrated scheme capable of constant re-evaluation. Until recently no satisfactory methods for such re-evaluation existed, but the advent of computer-based dynamic programming goes some way towards resolving the paradox. Buras [1966], in a very interesting paper on this subject, outlines the possibilities for greater use and development of dynamic programming, particularly with reference to conjunctive management of surface and ground-water storage. Although this approach faces many, as yet, unsolved problems (e.g. in flow probabilities and analysis of large systems), it seems that the most efficient choice in water use will only be made when spatial and dynamic aspects of the problem can be considered in detail simultaneously.

REFERENCES

BURAS, N. [1966], Dynamic programming in water resources development; in Chow, V. T., Editor, *Advances in Hydroscience*, Vol. 3 (Academic Press, New York and London).

BURTON, I. and KATES, R. W., Editors [1965], *Readings in Resource Management* (Chicago Univ. Press). (See especially chapters by Hufschmidt, M. M. and by Sewell, W. R. D. *et al.*)

DAHL, R. A. and LINDBLOM, C. E. [1953], *Politics, Economics and Welfare* (New York), 557 p.

ECKSTEIN, O. [1958], *Water Resource Development: The economics of project evaluation* (Cambridge, Mass.)

FOX, I. K. [1966], We can solve our water problems; *Water Resources Research*, **2**, 617–23.

HIRSCHLEIFER, J., DE HAVEN, J. C., and MILLIMAN, J. W. [1960], *Water Supply; Economics, Technology and Policy* (Chicago), 378 p.

HUFSCHMIDT, M. M. and FIERING, M. B. [1966], *Simulation Techniques for Design of Water Resource Systems* (Harvard University Press), 212 p.

KATES, R. W. [1962], *Hazard and Choice Perception in Flood Plain Management*; University of Chicago, Department of Geography Research Paper No. 78, 157 p.

KRUTILLA, J. V. and ECKSTEIN, O. [1958], *Multiple Purpose River Development* (Baltimore), 301 p.

KRUTILLA, J. V. [1967], *The Columbia River Treaty – The Economics of an International River Basin Development* (Baltimore), 211 p.

KUIPER, E. [1965], *Water Resources Development, Planning, Engineering and Economics* (London), 483 p.

LOWENTHAL, D., Editor [1966], *Environmental Perception and Behavior;* University of Chicago, Department of Geography, Research Paper No. 109, 88 p.

MAASS, A., HUFSCHMIDT, M. M., DORFMAN, R., THOMAS, H. A., MARGLAN, S. A. and FAIR, G. M. [1962], *Design of Water-Resource Systems* (Cambridge, Mass.), 620 p.

MCKEAN, R. N. [1958], *Efficiency in Government through Systems Analysis: With Emphasis on Water Resources Development* (New York), 336 p.

MORE, R. J. [1967], Hydrological models and geography; In Chorley, R. J. and Haggett, P., editors, *Models in Geography* (London), pp. 145–85.

NATIONAL ACADEMY OF SCIENCES [1966], Alternatives in Water Management; *National Research Council Committee on Water, Publication No.* 1408, (Washington, D.C.).

O'RIORDAN, T. and MORE, R. J. [1969], Choice in water use; *Transactions of the Institution of Water Engineers.*

PREST, A. R. and TURVEY, R. [1966], Cost–Benefit Analysis – a Survey; *Economic Journal*, **16**, 683–735.

SAARINEN, T. F. [1966], *Perception of the Drought Hazard in the Great Plains*; University of Chicago, Department of Geography Research Paper No. 106, 198 p.

SEWELL, W. R. D., DAVIS, J., SCOTT, A. D. and ROSS, D. W. [1961], Guide to benefit-cost analysis; *Resources for Tomorrow Conference, Background Papers I*, (Queen's Printer, Ottawa), p. 17.

SIMON, H. A. [1957], *Models of Man* (New York), 287 p.

SPARGO, R. A. [1961], Benefit Cost Analysis and Project Evaluation; *Resources for Tomorrow Conference, Background Papers* 1 (Queen's Printer, Ottawa), pp. 299–310.

UDALL, S. C. [1963], *The Quiet Crisis* (New York), 224 p. (See especially pp. 132–4.)

WATER RESOURCES BOARD [1966], *Water Supplies in South-East England* (HMSO, London).

WENGERT, N. [1955], *Natural Resources and the Political Struggle* (New York), 71 p.

WHITE, G. F. [1957], Trends in River Basin Development; *Law and Contemporary Affairs*, **22**, 157–87.

WHITE, G. F. [1964], *Choice of Adjustment to Floods*; University of Chicago, Department of Geography Research Paper No. 93, 164 p.

Index